ACTIVE
PHYTOCHEMICALS
FROM CHINESE
HERBAL
MEDICINES

ACTIVE PHYTOCHEMICALS FROM CHINESE HERBAL MEDICINES

ANTI-CANCER ACTIVITIES AND MECHANISMS

WING SHING HO

CHINESE UNIVERSITY OF HONG KONG, CHINA

CRC Press
Taylor & Francis Group
Boca Raton London New York

CRC Press is an imprint of the
Taylor & Francis Group, an **informa** business

CRC Press
Taylor & Francis Group
6000 Broken Sound Parkway NW, Suite 300
Boca Raton, FL 33487-2742

First issued in paperback 2017

© 2016 by Taylor & Francis Group, LLC
CRC Press is an imprint of Taylor & Francis Group, an Informa business

No claim to original U.S. Government works

ISBN-13: 978-1-4822-1986-9 (hbk)
ISBN-13: 978-1-138-89439-6 (pbk)

Visit the Taylor & Francis Web site at
http://www.taylorandfrancis.com

and the CRC Press Web site at
http://www.crcpress.com

I dedicate this book to my family: my wife, Teresa, my two children, Yuen and Hang, and the memories of my mother, Tong Sam Mui, and of my father, Ho Tak Kam, who helped me with my career and bringing up my children.

Contents

Preface

This is the first book on integrated pharmacology of herbal medicines with a unique approach toward the development of phytochemicals and their mechanisms of actions in the context of cancers and the diseases they are used to treat. The book covers biologic actions of the active phytochemicals at the molecular, cellular, and organ levels. The first few chapters deal with the principles of the interaction of phytochemicals and the related drug actions. The book also covers the basic concepts of identification of active phytochemicals and their pharmacological actions. It provides insightful information on how our knowledge can be influenced by the biologic and chemical factors of phytochemicals. Conventional icons are used to explain the main molecular and cellular actions of phytochemicals for better understanding. These icons deal with the treatment of cancer and diseases treated with herbal phytochemicals in rats. Each body system addresses the common pathways affecting cancer development before discussing the phytochemical classes and specific phytochemicals that have been recently reported in journal papers for the management of cancer and other diseases. The readers are introduced to the increasingly important aspects of pharmacology, including health benefits and drawbacks of phytochemicals. In addition, relevant background of the biochemistry of cancer is provided. Necessary illustrations that depict relevant pharmacology that would enhance the understanding of phytochemical actions are included. Tables presenting their adverse actions are used to highlight important issues related to phytochemical actions. Prescription drugs are used to compare phytochemical actions in in vivo studies. Insights into how phytochemicals can be developed from herbal medicines with multiarrays of pharmacological activities are offered. We hope that this book will provide useful information and reference on phytochemicals obtained from herbal medicines and will contribute toward cancer drug development with herbal phytochemicals. It has been designed for medicinal scientists but will also be useful to pharmaceutical professionals and students interested in alternative medicines, as it bridges the gap between fundamental mechanisms of anticancer actions and the use of phytochemicals to manage cancers and other human diseases. We hope that the readers enjoy reading this book.

Author

Dr. Wing Shing Ho, PhD, is associate professor of the biochemistry programme at the School of Life Sciences in The Chinese University of Hong Kong, Hong Kong, China. He earned his BS in biochemistry (1979) from the University of Alberta and MA in chemistry (1982) and PhD in biological chemistry (1985) from the State University of New York at Buffalo. After completing a postdoctoral training in the pediatrics department at SUNYAB, he moved to the Department of Chemistry at the University of Utah as a postdoc investigating the methodology of isolation and purification of DNA and subsequently moved to the Center for Human Toxicology, University of Utah, as a research associate investigating the role of hepatic toxicants on liver metabolism in lab animals. In 1994, Dr. Ho was appointed lecturer in the Department of Biochemistry and, in 2005, he was appointed associate professor at The Chinese University of Hong Kong and became an instrumental part of the toxicology programme in the School of Life Sciences.

Dr. Ho holds memberships in several professional associations, including the United States Society of Toxicology, the American Chemical Society, the New York Academy of Sciences, the American Institute of Chemists (fellow), and the Protein Society. He has been appointed a consultant scientist by professional groups and the local government, including Government Secretariat Home Affairs Bureau and the HK Chemical Waste Association.

Dr. Ho's work has been supported in part by the university research grants from the Hong Kong Higher Education and Innovation and Technology Commission and the Croucher foundation. He has received awards from the American Society for Biochemistry and Molecular Biology for students' papers.

Dr. Ho has authored and coauthored approximately 100 papers and proceedings in peer-reviewed international journals and holds patents on herbal medicines. He has contributed original data to the Protein Data Bank (PDB) and has applied for patents under the Patent Cooperation Treaty (PCT) for anticancer agents in the United States. He lectures regularly on toxicology to undergraduate, and graduate students. Dr. Ho continues to perform fundamental research on the cytotoxic effects of environmental and food chemicals and the development of therapeutic agents on cultured human cells and animal models.

Abbreviations

Abbreviations	Terms
ADME	Absorption, distribution, metabolism, and excretion
AML	Acute myeloid leukemia
BM	*Bacopa monnieri*
CAM	Complementary and alternative medicine
CAT	Catalase
CI	Combination index
CIA	Collagen-induced arthritis
COX-1	Cyclooxygenase-1
COX-2	Cyclooxygenase-2
CRS	Chemical reference substances
DHA	Dihydroartemisinin
EBV-EA	Epstein-Barr virus early antigen
ER	Estrogen receptor
ERS	Extractive reference substance
FAHF	Food allergy herbal formula
GA	Gambogic acid
GABAA	Gamma-aminobutyric acid (Type A)
GSK3	Glycogen synthase 3 kinase
GT	Gallotannin
H1	Histamine type 1 receptor
HBV	Hepatitis B virus
HCV	Hepatitis C virus
HDL	High density lipoprotein
^{1}H NMR	Proton nuclear magnetic resonance
HO-1	Heme oxygenase-1
HSCs	Hepatocyte stellate cells
ICA	Independent component analysis
IKK	I kappa B kinase
IL	Interleukin
Iso-GNA	Isogambogenic acid
KPS	Karnofsky performance score
LCAT	Lecithin cholesterol acyltransferase
LDH	Lactate dehydrogenase
LPL	Lipoprotein lipase
MDR	Multidrug resistance
MFAX	Methanol fraction of amomum xanthoides

NF	Nuclear factor
NIH	National Institutes of Health
NSAID	Nonsteroid anti-inflammatory drug
PBMC	Peripheral blood mononuclear cells
PCA	Principal component analysis
PCNA	Proliferating cell nuclear antigen
PDE5	Phosphodiesterase (Type 5)
PI3K	Phosphatidylinositol-3-kinase
PKC	Protein kinase C
PMA	Phorbol myristate acetate
PPARs	Peroxisome proliferator-activated receptors
PTEN	Phosphatase and tensin homologue
PXR	Pregnane X receptor
RA	Rheumatoid arthritis
ROS	Reactive oxygen species
RR	Relative risk
RTKs	Receptor typosine kinases
SOD1	Superoxide dismutase 1
Tan I	Tanshinone I
TCE	Trichloroethylene
TCM	Traditional Chinese medicine
THL	Tien–Hsien Liquid
TI	Therapeutic index
TLBZT	Teng-Long-Bu-Zhong-Tang
TNF	Tumor necrosis factor
UGTs	UDP-glucuronosyltransferases
VEGF	Vascular endothelial growth factor

1 Introduction

There is an increasing use of herbal medicines for the treatment of various ailments worldwide. Herbal medicines are often taken in combination with other therapeutic drugs in China and other Asian countries. For thousands of years, specific herbal formulations have been used for the treatment of various diseases in China. It is believed that individual herbs in the formulations would act on different targets and systems to produce the expected health benefits. However, the combination of herbs with therapeutic drugs can raise potential health risk. The health risk is probably attributed to drug–herbal interactions. Data from human studies indicate that *Hypericum perforatum*, also known as St. John's wort, decreased the blood concentrations of amitriptyline, warfarin, and theophylline, whereas it did not alter the pharmacokinetics of dextromethorphan, mycophenolic acid, and pravastatin [1]. *H. perforatum* also decreased the plasma concentration of the active metabolite SN-38 in cancer patients after irinotecan treatment. Both pharmacokinetics and pharmacodynamics are believed to be affected by drugs–herbal interactions. The potential interactions of herbal active ingredients with drugs are of a major concern in the combination therapy. Administration of drugs and herbal medicines requires appropriate strategy in order to reduce potential herb–drug interactions, and to enhance herbal safety and efficacy.

HERBAL MEDICINES AS COMPLEMENTARY MEDICINE FOR CANCER THERAPY

In Chinese medicine, mortal damage is a consequence of the disorientation and imbalance of "Yin-Yang," which helps human beings try to sustain harmony and integrity. Cancers and tumors are the consequence of unmitigated accumulations of "qi," moisture, and blood that have intoxicated the system [2]. These factors are transformed into morbid tissue, obstructing the normal circulation of blood and lymph. Consequently, it leads to depletion of "qi" and blood resulting in a deregulation of growth typical of cancer. Therefore, treatment that supplements "qi," moisture, and blood can restore circulation and alternates stasis. This can replenish the essence that governs growth and maturation and repair mechanism.

It is believed that the relationship between generating blood and its circulation is meant to prevent both deficiency and stagnation. The blood flows to every part of the body and moistens and lubricates all the tissues.

When there is insufficient "ying qi," the distribution of "qi" is significantly interrupted. Consequently, therapeutic measures are administered to remove the obstruction and generate new blood. Chinese herbs such as angelica, salvia, and millettia can be used to treat both deficiency and stasis because they can enhance blood circulation.

Cancer patterns involve accumulation of carcinogenic metabolites, deficient "qi" and blood, and blood stagnation. For example, stomach cancer may result from various patterns such as liver "qi" invading the stomach, stomach "yang" deficiency, phlegm stagnation and blood stagnation due to "qi" stagnation, stomach "yin" deficiency due to stomach heat, and "qi" and blood deficiency. Depending on the patterns of deficiency, herbal therapies are individualized to fit different patients according to the pathological patterns and the health status of the patient. Herbal formulas relieve stagnation by using "qi-" and blood-activating herbs and antidote toxins. Anticancer herbs can remove or dissolve these pathogenic entities.

There are many treatment strategies that combine acupuncture and Chinese medicines to reduce toxins and enhance the reduced circulation, both of which are caused by tumors, cancers, and the adverse effects of surgery, radiation, and chemotherapy. Herbal formulas are especially effective in invigorating the "qi," nourishing the blood, and clearing the system. Herbal medicines can enhance adaptation to stress and increase host defense mechanism, which will increase resistance to inflammation and infection and suppress the progression of tumors. Consequently, herbal medicines can extend life span of patients. These are primary therapeutic protocols for the management of cancer in China and some other Asian countries. Chinese herbal medicine represents complementary or adjunctive therapies that often can improve the efficacy of Western medicine to achieve the pharmacological effects.

The signs and symptoms of deficiency of "qi," moisture, and blood manifest themselves as the adverse effects of radiation and chemotherapy. Replenishing these deficiencies of "qi," moisture, and blood require medication. Herbal medicines become the top choice especially in generating these resources in the system. While Western medicine aggressively inhibits the cancer, yet, with side effects, traditional Chinese medicine supports and restores the functions of vital organs that enable patients to recover slowly from conventional therapies. The quality of life of patients can be improved.

MODERN CHINESE HERBAL RESEARCH

With the advent of science and biotechnology, clinical researchers worldwide, especially from China and Japan, have begun searching for ways to improve treatment strategy in combination with herbal medicine.

Over the last decades, this approach has become known as Fuzheng Gu Ben therapy, which is meant to strengthen the existing health status and secure the root. Fuzheng herbs support nonspecific resistance and are also known as adaptogens. It is believed that patients with advanced malignant tumors usually show symptoms of deficiency in "qi," blood, liver, and kidney, as well as dysfunction of spleen and stomach. Tonics may improve the general health condition and the immune function [3]. Tonic treatment benefits the treatment of cancer patients. Fuzheng therapy produces diverse biologic effects that include the following:

1. Reduction of tumor mass
2. Suppression of the formation of a new primary cancer
3. Enhancement of the immune function
4. Bolstering of the regulatory function of the endocrine system
5. Protection of the structure and function of internal organs and glands
6. Strengthening of the digestive system by improving absorption and metabolism
7. Protection of the bone marrow and hematopoietic function
8. Treatment of adverse side effects due to drug toxicity

A couple of excellent resources on the role of Chinese herbs in cancer therapy is *Cancer and Natural Medicine* and *Natural Compounds in Cancer Therapy*, which are authored by John Boik [4,5].

Herbal medicines have become increasingly popular among cancer patients. They often use herbal medicines as adjuvant therapy to reduce the side effects of the common chemotherapy. However, interactions between herbal medicines and cancer drugs can occur if administered inappropriately. These herb–cancer drug interactions can reduce the efficacy of cancer drugs and cause adverse consequences for cancer patients. Adverse interactions between herbal medicines and therapeutic drugs have been reported. The herb–drug interactions are involved with drug-metabolizing enzyme system. Extracts of *Oldenlandia diffusa* and *Astragalus propinquus* were reported to inhibit CYP3A4 in human [6]; these herbs significantly induced human pregnane X receptor (PXR)-mediated CYP3A4. Concomitant use of *O. diffusa* and *Rehmannia glutinosa* resulted in the induction of CYP3A4 and consequently reduced the efficacy of drugs. Herb–drug interactions between herbal medicines and CYP3A4 substrates can occur.

Despite potential health risk associated with herb–cancer drugs interactions, cancer patients always take herbal supplements in order to reduce cytotoxicity of cancer drugs. In women with advanced breast cancer, coadministration of garlic supplement reduced the clearance of docetaxel [7].

The study did not alleviate cancer drugs–induced hematological toxicity, but alternate drug–induced nausea. Most cancer drugs are substrates of P-glycoprotein and can cause multidrug resistance (MDR)–associated proteins and other transporters [7]. Induction and inhibition of the drug-metabolizing enzymes and transporters may affect efficacy of therapeutic drugs. Therefore, the selection of appropriate herbal medicines and the cancer drug for cancer therapy is important to avoid health hazard.

HEALTH BENEFITS OF HERBAL MEDICINES

A large number of herbal medicines are used for treating cancer or reducing the cytotoxicity of chemotherapeutic drugs. Some of the herbal medicines have been reported to show health effects on cancer and attenuate therapeutic drugs–induced toxicities. However, there is not much clinical data associated with the use of herbal medicines in cancer therapy. A survey has identified herb–chemotherapeutic drug combinations in the selected group of cancer patients [8]. Among 42 cancer patients using herbal medicinal remedies in combination with chemotherapy, 47 different potential herb–drug interactions were identified on the level of cytochrome P-450 metabolizing enzymes and glycoprotein transport in vitro. Common herbal remedies included garlic, ginger, and green tea. The clinical potential for cytochrome P-450 metabolizing enzyme interactions in humans was reported for green tea and echinacea. The study reported that garlic displayed strong interactions with glycoprotein. In addition, the use of food supplements including practice among cancer patients. The concurrent use of complementary medicine and natural health products with therapeutic drugs may reduce the efficacy of radiotherapy as well.

Although conventional treatments with cancer drugs can extend the life of cancer patients, the adverse side effects of these cancer drugs pose a limitation to cancer therapy. The cytotoxicity causes significant psychological and spiritual stress in cancer patients. Neither herbal medicines nor Western medicine alone can satisfactorily alleviate the stress associated with therapeutic drugs. Herbal medicine modalities offer less toxicity; yet, effective treatment for advanced cancers remains to be investigated. A combination of Western medicine and herbal medicines would complement each other for the treatment of cancer. Health risks of integrating herbal medicines with cancer drugs can be ameliorated. An integrative approach can harness the strengths of Western medicine and herbal medicine [9].

Cancer and the related complications significantly compromise immune response and the quality of life. The pharmacological effects of herbal medicines such as anticancer medicines or adjuvants can increase the cancer drug efficacy and ameliorate undesirable side effects of cancer drugs.

The combination of drugs and herbal medicines revealed that the antitumor immunity can be improved [10].

HERBAL MEDICINES AS A RICH SOURCE FOR DRUG DEVELOPMENT

Herbal medicine has become a rich source of anticancer agents and facilitates the development of efficacious cancer drugs. However, extensive screening tests of the potential active phytochemicals in vitro are needed. In vitro combination and characterization of potential anticancer phytochemicals against human cancer cell lines are commonly performed. Potentially effective combinations of beta-elemene with taxanes were explored and demonstrated in human lung cancer cells [11]. Synergistic interactions were observed with combinations of ss-elemene and taxanes. This is related to the enhanced cytotoxicity of taxanes via induction of cytochrome c release from mitochondria, caspase-8, and -3 cleavage and downregulation of Bcl-2 and Bcl-X-L expression. The anticancer activity of the combination of herbal medicines and cancer drugs has raised tremendous interest worldwide.

The health benefits of herbal medicines are attributed to the synergistic interactions with anticancer compounds. This combination strategy for the treatment of cancer is used to evaluate the composition of traditional Chinese medicine formulation. Yanhusuo San consisting of Rhizoma Corydalis and Rhizoma Curcumae was an ancient Chinese medicine prescription for the treatment of cancer dated back in AD 960–1279 [12]. A common approach is to compare the IC_{50} of each herbal extract and both extracts at different compositions by MTT assay. The isobologram and combination index (CI) are used to evaluate the synergistic effects of the herbal extracts. Flow cytometry, fluorescence analysis, Western blot analysis, and gene expression profile can be used to fathom out the mechanism of actions of the herbal extracts. A plausible molecular mechanism of the synergistic antitumor effects of Rhizoma Corydalis and Rhizoma Curcumae was reported [11].

ACTIVITY-BASED FRACTIONATION OF HERBAL EXTRACTS

The active fractions are derived from the activity-based fractionation of herbal extracts in in vitro and in vivo study. However, the antitumor activity of single phytochemicals may not show effects in animal study. Nevertheless, the composition of phytochemicals with antitumor effects in cancer cell lines can be evaluated in vivo. *Curcuma aromatica* and *Polygonum cuspidatum* are widely used herbs for liver cancer therapy. Curcumin, the active principle of *C.aromatica*, and resveratrol, the active principle of *P. cuspidatum*, contribute to the anticancer effects in colon cancer [13]. The combination of curcumin and resveratrol significantly inhibited the proliferation of Hepa-1-6 cells. The combination of phytochemicals may enhance the efficacy of herbal medicine. The combination of curcumin and resveratrol is a good example of combination strategy for liver cancer treatment.

Current data basis of herbal medicine is limited. Bioinformatics on the selection and synergism of herbal extracts and constituents with anticancer properties are warranted to be developed. Chemoinformatics methodology can play an important role in clinical applications and are helpful in drug development. Novel herbal extract in combination of cancer drugs were characterized in ACHN and A2780/cp cells with chemobioinformatics-aided analysis [14]. Chemobioinformatics confirmed the predicted outcomes. It could provide useful information on the use of herbs in reversing MDR.

Herbal medicines are commonly used in food supplement for improving general health. Diet is not the primary therapy for refractory forms of cancer. Yet an appropriate food supplement may be effective as an adjuvant to hormone deprivation therapy for cancer. This treatment strategy could delay relapse and inhibit refractory growth. Zyflamend, a combination of multiple herbal extracts, was reported to exhibit anticancer properties [15,16]. Zyflamend can inhibit growth of various prostate cancer cell lines and androgen-dependent tumor growth in a mouse model at the advanced stages of prostate cancer. Herbal extracts as adjuvant in food supplements have become a common approach in cancer therapy. The active herbal extracts also exhibit chemoprotective properties against carcinogenesis.

Herbal medicines generally show health benefits toward different types of cancer. They have been used as complementary and alternative medicine for different cancer therapies. Although herbal medicines may take a longer period of time to show pharmacological effects in cancer patients, the adverse side effects of herbal medicines are minimal. Often, herbal formulations are preferred based on clinical practices. Teng-Long-Bu-Zhong-Tang (TLBZT) was demonstrated to show anticancer effects on colorectal cancer in vitro [17]. TLBZT significantly inhibited CT26 colon carcinoma growth via apoptotic caspase cascade. It can enhance the anticancer effects of 5-Fu in CT26 colon carcinoma.

Herbal medicines are known to be able to enhance the efficacy of cancer drugs. A combination of the active constituents can be more effective anticancer agents. A combination of Jaceosidin, emodin, and magnolol showed remarkable anticancer activities in melanoma A375 [18]. The combination of these phytochemicals induced cell cycle arrest and, consequently, apoptosis in melanoma A375 cell line. However, a different combination of emodin with magnolol was more effective than the other combinations.

TRADITIONAL HERBAL FORMULATIONS

Traditional herbal formulations represent specific composition of herbal extracts prepared in boiling water. However, it is realized that temperature effect may have a significant impact on extraction efficiency and stability of phytochemicals [19,20]. The chemical properties of chlorogenic acid, an active component in green tea and traditional Chinese medicine, can be changed through boiling, yet, the anticancer properties of chlorogenic acid and its derivatives remained the same against CCl_4-induced toxicity in hepG2 cells [19]. Herbal extraction of *Scutellaria baicalensis* Georgi, *Glycyrrhiza uralensis* Fisch, *Paeonia lactiflora* Pall, and *Ziziphus jujuba* Mill in boiling water enhanced the anticancer activities of chemotherapy in various cancers in various mouse tumor xenograft and allograft models [20].

CLINICAL EVALUATION OF HERBAL MEDICINES

Laboratory studies of herbal medicines are expanding the clinical knowledge. Different herbal extracts can be prepared as natural health products. Specific formulations are more effective to target molecular pathways, including angiogenesis and epidermal growth factor receptor, which plays a significant role in cancer growth. Quality assurance of specific herbal extracts is important as herbs obtained from different geographical regions might not have the same anticancer properties. Their effectiveness may be affected when multiple herbal agents are used. An integrative approach for managing cancer should target the multiple signaling pathways associated with cancer growth. Angiogenesis is an essential process in cancer development. Herbal products can affect angiogenesis and consequently inhibit cancer growth [21]; thus, herbal extracts with antiangiogenic activity may be good anticancer agents. Examples of antiangiogenic herbs include *Artemisia annua*, *Viscum album*, *Curcuma longa*, *S. baicalensis*, *Magnolia officinalis*, *Camellia sinensis*, *Ginkgo biloba*, *Panax ginseng*, and quercetin. Surprisingly, other active phytochemicals may not show high anticancer effects when used alone, yet they can interact with other components in the herbal extracts to produce anticancer activities. Baicalin, an active phytochemical with antipyretic properties, in combination with *Salvia miltiorrhiza* or *C. sinensis* extracts showed antiproliferation effects on the human breast cancer cell lines MCF-7 and T-47D [22]. The combination of active compounds from different classes offers either enhanced therapeutic benefits or antiproliferative effects on tumor growth. The antiproliferative effects of these compounds can be extended to other cancer types, suggesting that these compounds in combination with herbal extracts may function through different mechanisms. The advantages of using herbal medicines in cancer therapy are demonstrated in different studies either in combination with cancer drugs or in combination with other herbal extracts.

Increasing failure rates with cancer chemotherapy, the high cost, and limited drug efficacy have prompted alternative approaches to cancer treatment and drug discovery. Common molecular basis of cancer biology has been discovered through advancements in genomics and proteomics. With the advent of biotechnologies, a large numbers of potential drug candidates can be tested against a particular molecular target; thus, novel cancer drugs derived from herbal principles can be developed. One of the most abundant natural sources for active compounds is herbal medicines. Natural products and their derivatives have been purified and structurally identified from herbal medicines. These active phytochemicals exhibit immense pharmacological and anticancer properties. Although the molecular mechanism of actions of active anticancer phytochemicals are yet to be elucidated, extensive research in herbal medicine continues to generate new data that are worth further investigation in clinical testing. Recent advancement in biotechnology and chemical technology has enabled us to understand the salient interactions of natural products and their derivatives with cancer cells. As a result, the findings allow us to better design cancer drugs. Both the natural products and synthetic molecules share high chemical selectivity and pharmacological specificity. The ability of novel natural products to interact with specific protein domains would trigger a chain of cellular signaling processes. By virtue of specific binding affinity to gene products, natural products can provide effective scaffolds for the pharmacological processes [23]. Potent natural products should show properties specified in the "Lipinski's rule of five" [24]. The application of bioinformatics to quality control and the efficacy of natural product derivatives, in particular small natural molecules, should be used to enhance the interactions with molecular targets. However, these small naturally occurring products may involve multiple cellular targets and pathways. The inhibition of multiple mechanistic pathways may ameliorate chemoresistance in various cancers. Table 1.1 shows the anticancer properties of active phytochemicals and their mechanism of actions. A better understanding of the interactions between natural compounds and the derivatives in cancer cells is pivotal for the development of targeted anticancer agents. The discovery and development of potent anticancer agents from herbal medicines is hampered by lack of preclinical model for the evaluation of the potency of anticancer phytochemicals. However, xenograft models and transgenetic models can serve as surrogates in developing therapeutic anticancer drugs. Nevertheless, these models can show significant biochemical and physiological difference from humans. One of the potential issues associated with animal models for evaluation of potential drug efficacy is inappropriate dosing. With the

TABLE 1.1
Anticancer Activities of Active Phytochemicals

Herbal Compound	Specialty	References
Berberine	1. Completely antagonizes the TNF alpha-mediated barrier defects in the cell model 2. Dysregulation of protein folding, proteolysis, redox regulation, protein trafficking, cell signaling, electron transport, metabolism, and centrosomal structure in breast cancer cells	[26,27]
Baicalein	1. Inhibits the invasion of MDA-MB-231 human breast cancer cells 2. Inhibits tumor cell invasion and metastasis by reducing cell motility and migration via the suppression of the extracellular signal-regulated kinases (ERK) pathway	[28,29]
Rocaglamide	1. Sensitizes CD95L- and TRAIL-induced apoptosis in HTLV-1-infected cells by downregulation of c-FLIP expression 2. Blocks DNA damage-induced upregulation of the transcription factor p53 by inhibiting its protein synthesis	[30,31]
Shikonin	Inhibits tumor cell growth in estrogen receptor alpha (ER alpha)-positive, but not ER alpha-negative breast cancer cells	[32]
Rhein	Dysregulation of cytoskeleton regulation, protein folding, glycolysis pathway, and transcription control in breast cancer cells	[33]
Silymarin	1. Membrane-stabilizing and antioxidant activity promotes hepatocyte regeneration 2. Reduces the inflammatory reaction, inhibits the fibrogenesis in the liver	[34,35]
Catechin hydrate	Antiproliferative effects of CH in the prevention of cervical cancer	[36]
Ursolic and oleanolic acids	1. Upregulates p53 expression and inhibits breast cancer cell growth 2. Exerts antiproliferative and apoptotic effects selectively in ERa-positive breast cancer cells	[37]
Tanshinone I (Tan I)	Induction of apoptosis by Tan I in leukemia cells	[38]
Dihydroartemisinin (DHA)	Induces cell apoptosis by triggering reactive oxygen species (ROS)-mediated caspase-8/Bid activation and the mitochondrial pathway	[39]

(Continued)

TABLE 1.1 (*Continued*)
Anticancer Activities of Active Phytochemicals

Herbal Compound	Specialty	References
Gallotannin (GT)	Downregulates the expression of NF kappa B-regulated inflammatory cytokines (IL-8, TNF alpha, IL-1 alpha) and caused cell cycle arrest and accumulation of cells in pre-G(1) phase	[40]
Oridonin	Apoptosis- and autophagy-inducing activity in cancer therapy	[41]
Fei-Liu-Ping ointment	Reduces the concentration of serum proinflammatory cytokines IL-6, TNF-[alpha], and IL-1[beta]	[42]
Isogambogenic acid (iso-GNA)	Inhibits tumor angiogenesis	[43]

improvement of animal models for human tumors at different developing stages, anticancer activities of potent phytochemicals that are identified from herbal extracts are likely to be helpful in drug development [25].

Identification of specific biomarkers in cancer therapeutics is essential for the comparison of efficacy of potential phytochemicals from herbal extracts. Bioinformatics on different biological systems offers useful information related to molecular signaling activity and assays for cancer biomarkers.

The use of herbal formulation for cancer therapy depends on the documented ancient formula and the quality of herbs that can vary significantly in phytochemical contents and anticancer properties due to geographical differences in herbal medicines. This has hampered the clinical use of ancient herbal formulation for the treatment of cancer. Establishment of a useful relationship between a potent anticancer photochemical and cancer remains a challenge. Although cancer biology has been helpful in fathoming out the pathogenesis, only a limited number of the preventive measures based on herbal medicines are successful in human. However, herbal medicines remain one of the rich resources of anticancer agents for novel drug development. A combination strategy with cancer drugs and herbal medicines should be established for cancer therapy.

REFERENCES

1. Zhou, S., Chan, E., Pan, S.Q., Huang, M., and Lee, E.J. 2004. Pharmacokinetic interactions of drug with St John's Wort. *J. Psychopharmacol.* 18: 262–272.
2. Benefield, H. and Korngold, E. 2003. Chinese medicine and cancer care. *Altern. Ther. Health Med.* 9(5): 38–52.

3. Guorui, T. 1981. *The Use of Tonics in China—Past, Present, and Future.* Academy of Traditional Chinese Medicine, Beijing, China.

4. Boik, J. 1995. *Cancer and Natural Medicine.* Oregon Medical Press, Princeton, MN.

5. Boik, J. 2001. *Natural Compounds in Cancer Therapy.* Oregon Medical Press, Princeton, MN. www.ompress.com.

6. Lau, C., Mooiman, K.D., Maas-Bakker, R.F., Beijnen, J.H., Schellens, J.H., and Meijerman, I. 2013. Effect of Chinese herbs on CYP3A4 activity and expression in vitro. *J. Ethnopharmacol.* 149: 543–549.

7. Yang, A.K., He, S.M., Liu, L., Liu, J.P., Wei, M.Q., and Zhou, S.F. 2010. Herbal interactions with anticancer drugs: Mechanistic and clinical considerations. *Curr. Med. Chem.* 17: 1635–1678.

8. Engdal, S., Klepp, O., and Nilsen, O.G. 2009. Identification and exploration of herb-drug combinations used by cancer patients. *Integr. Cancer Ther.* 8: 29–36.

9. Li, P., Chen, J., Wang, J., Zhou, W., Wang, X., Li, B., Tao, W., Wang, W., Wang, Y., and Yang, L. 2014. Systems pharmacology strategies for drug discovery and combination with applications to cardiovascular diseases. *J. Ethnopharmacol.* 151(1): 93–107.

10. Wang, H., Chan, Y.L., Li, T.L., and Wu, C.J. 2012. Improving cachectic symptoms and immune strength of tumor-bearing mice in chemotherapy by a combination of *Scutellaria baicalensis* and Qing-Shu-yi-Qi-Tang. *Eur. J. Cancer* 48: 1074–1084.

11. Zhao, J., Li, Q.Q., Zou, B., Wang, G., Li, X., Kim, J.E., Cuff, C.F., Huang, L., Read, E., and Gardner, K. 2007. In vitro combination characterization of the new anticancer plant drug beta-elemene with taxanes against human lung carcinoma. *Int. J. Oncol.* 31: 241–252.

12. Gao, J.L., He, T.C., Li, Y.B., and Wang, Y.T. 2009. A traditional Chinese medicine formulation consisting of Rhizoma Corydalis and Rhizoma Curcumae exerts synergistic anti-tumor activity. *Oncol. Rep.* 22: 1077–1083.

13. Du, Q., Hu, B., An, H.M., Shen, K.P., Xu, L., Deng, S., and Wei, M.M. 2013. Synergistic anticancer effects of curcumin and resveratrol in Hepal-6 hepatocellular carcinoma cells. *Oncol. Rep.* 29: 1851–1858.

14. Ghavami, G., Sardari, S., and Shokrgozar, M.A. 2011. Chemoinformatics-based selection and synergism of herbal extracts with anti-cancer agents on drug resistance tumor cells-ACHN and A2780/CP cell lines. *Comput. Biol. Med.* 41: 665–674.

15. Huang, E.C., McEntee, M.F., and Whelan, J. 2012. Zyflamend, a combination of herbal extracts, attenuates tumor growth in murins xenograft models of prostate cancer. *Nutr. Cancer Int. J.* 64: 749–760.

16. Park, K.W., Ye, S.H., Kim, Y.J., Jang, S.R., Bang, M.H., Lee, H.W., and Park, K.M. 2010. In vitro and in vivo anti-tumor effects of oriental herbal mixtures. *Food Sci. Biotechnol.* 19: 1019–1027.

17. Deng, S., Hu, B., An, H.M., Du, Q., Xu, L., Shen, K.P., Shi, X.F., Wei, M.M., and Wu, Y. 2013. Teng-Long-Bu-Zhong-Tang, a Chinese herbal formula, enhances anticancer effects of 5-Fluorouracil in CT26 colon carcinoma. *BMC Complement. Altern. Med.* 13: 128–134.

18. Shawi, A., Kimatu, J.N., Khan, M., and Hussain, K.A. 2011. Enhancement of induced apoptosis in human melanoma A375 by a combination of natural compounds. *J. Med. Plants Res.* 5: 5400–5406.

19. Kan, S., Cheung, M.W.M., Zhou, Y., and Ho, W.S. 2013. Effects of boiling on chlorogenic acid and the liver protective effects of its main products against CCl4-induced toxicity in vitro. *J. Food Sci.* 79(2): c147–c154.

20. Liu, S.H. and Cheng, Y.C. 2012. Old formula, new Rx: The journey of PHY906 as cancer adjuvant therapy. *J. Ethnopharmacol.* 140: 614–623.

21. Sagar, S.M., Yance, D., and Wong, R.K. 2006. Natural health products that inhibit angiogenesis: A potential source for investigational new agents to treat cancer. *Curr. Oncol.* 1: 14–26.

22. Franek, K.J., Zhou, Z., Zhang, W.D., and Chen, W.Y. 2005. In vitro studies of baicalin alone or in combination with *Salvia miltiorrhiza* extract as a potential anti-cancer agent. *Int. J. Oncol.* 26: 217–224.

23. Peczuh, M.W. and Hamilton, A.D. 2000. Peptide and protein recognition by designed molecules. *Chem. Rev.* 100: 2479–2494.

24. Lipinski, C.A., Lombardo, F., Dominy, B.W., and Feeney, P.J. 1997. Experimental and computational approaches to estimate solubility and permeability in drug discovery and developmental settings. *Adv. Drug Discov. Rev.* 23: 3–25.

25. Gopinathan, A. and Tuveson, D.A. 2008. The use of GEM models for experimental cancer therapeutics. *Dis. Model Mech.* 2: 83–86.

26. Amasheh, M., Fromm, A., Krug, S.M., Amasheh, S., Andres, S., Zeitz, M., Fromm, M., and Schulzke, J.D. 2010. TNF alpha-induced and berberine-antagonized tight junction barrier impairment via tyrosine kinase, Akt and NF kappa B signaling. *J. Cell Sci.* 123: 4145–4155.

27. Chou, H.C., Lu, Y.C., Cheng, C.S., Chen, Y.W., Lyu, P.C., Lin, C.W., Timms, J.F., and Chan, H.L. 2012. Proteomic and redox-proteomic analysis of berberine-induced cytotoxicity in breast cancer cells. *J. Proteom.* 75: 3158–3176.

28. Wang, L., Ling, Y., Chen, Y., Li, C.L., Feng, F., You, Q.D., Lu, N., and Guo, Q.L. 2010. Flavonoid baicalein suppresses adhesion, migration and invasion of MDA-MB-231 human breast cancer cells. *Cancer Lett.* 297: 42–48.

29. Chen, K.L., Zhang, S., Ji, Y.Y., Li, J., An, P., Ren, H.T., Liang, R.R., Yang, J., and Li, Z.F. 2013. Baicalein inhibits the invasion and metastatic capabilities of hepatocellular carcinoma cells via down-regulation of the ERK pathway. *PLoS One* 8: e72927.

30. Bleumink, M., Kohler, R., Giaisi, M., Proksch, P., Krammer, P.H., and Li-Weber, M. 2011. Rocaglamide breaks TRAIL resistance in HTLV-1-associated adult T-cell leukemia/lymphoma by translational suppression of c-FLIP expression. *Cell Death Differ.* 18: 362–370.

31. Becker, M.S., Schmezer, P., Breuer, R., Haas, S.F., Essers, M.A., Krammer, P.H., and Li-Weber, M. 2014. The traditional Chinese medical compound Rocaglamide protects nonmalignant primary cells from DNA damage-induced toxicity by inhibition of p53 expression. *Cell Death Dis.* 5: e1000.

32. Yao, Y. and Zhou, Q. 2010. A novel antiestrogen agent Shikonin inhibits estrogen-dependent gene transcription in human breast cancer cells. *Breast Cancer Res. Treat.* 121: 233–240.

33. Huang, H.J., Lin, C.C., Chou, H.C., Chen, Y.W., Lin, S.T., Lin, Y.C., Lin, D.Y., Lyu, K.W., and Chan, H.L. 2014. Proteomic analysis of rhein-induced cyt: ER stress mediates cell death in breast cancer cells. *Mol Biosyst.* 10: 3086–3100.

34. Feher, J. and Lengyel, G. 2012. Silymarin in the prevention and treatment of liver diseases and primary liver cancer. *Curr. Pharm. Biotechnol.* 13: 210–217.

35. Shaker, E., Mahmoud, H., and Mnaa, S. 2010. Silymarin, the antioxidant component and *Silybum marianum* extracts prevent liver damage. *Food Chem. Toxicol.* 48: 803–806.

36. Al-Hazzani, A.A. and Alshatwi, A.A. 2011. Catechin hydrate inhibits proliferation and mediates apoptosis of SiHa human cervical cancer cells. *Food Chem. Toxicol.* 49: 3281–3286.

37. Gu, G.W., Barone, I., Gelsomino, L., Giordano, C., Bonofiglio, D., Statti, G., Menichini, F., Catalano, S., and Ando, S. 2012. *Oldenlandia diffusa* extracts exert antiproliferative and apoptotic effects on human breast cancer cells through ER alpha/Sp1-mediated p53 activation. *J. Cell. Physiol.* 227: 3363–3372.

38. Liu, J.J., Liu, W.D., Yang, H.Z., Zhang, Y., Fang, Z.G., Liu, P.Q., Lin, D.J. et al. 2010. Inactivation of PI3k/Akt signaling pathway and activation of caspase-3 are involved in tanshinone I-induced apoptosis in myeloid leukemia cells in vitro. *Ann. Hematol.* 89: 1089–1097.

39. Lu, Y.Y., Chen, T.S., Wang, X.P., and Li, L. 2010. Single-cell analysis of dihydroartemisinin-induced apoptosis through reactive oxygen species-mediated caspase-8 activation and mitochondrial pathway in ASTC-a-1 cells using fluorescence imaging techniques. *J. Biomed. Opt.* 15: 046028.

40. Al-Halabi, R., Chedid, M.B., Abou Merhi, R., El-Hajj, H., Zahr, H., Schneider-Stock, R., Bazarbachi, A., and Gali-Muhtasib, H. 2011. Gallotannin inhibits NF kappa B signaling and growth of human colon cancer xenografts. *Cancer Biol. Ther.* 12: 59–68.

41. Liu, Z., Ouyang, L., Peng, H., and Zhang, W.Z. 2012. Oridonin: Targeting programmed cell death pathways as an anti-tumour agent. *Cell Prolif.* 45: 499–507.

42. Li, W., Chen, C., Saud, S. M., Geng, L., Zhang, G., Liu, R., and Hua, B. 2014. Fei-Liu-Ping ointment inhibits lung cancer growth and invasion by suppressing tumor inflammatory microenvironment. *BMC Complement. Altern. Med.* 14: 153.

43. Fan, Y., Peng, A., He, S., Shao, X., Nie, C., and Chen, L. 2013. Isogambogenic acid inhibits tumour angiogenesis by suppressing rho GTPases and vascular endothelial growth factor receptor 2 signalling pathway. *J. Chemother.* 25: 298–308.

2 Overview
General Principles

Herbal medicine emerged as a science when voluminous research papers revealed its therapeutic benefits in recent decade. The emphasis began to shift from describing what therapeutic drugs do to explaining how herbal medicine can work in enhancing the efficacy of drugs. In this chapter, we start out some general principles underlying the interaction of drugs with living systems. The interaction between drugs and cells is often through interaction with the molecular targets, or the drug receptor interaction; yet, we are far from the holy grail of being able to predict the pharmacological effects of a novel phytochemical substance, or to design a more potent chemical to produce a desired therapeutic effect. Nevertheless, we can identify general principles involved in herbal compounds.

INTERACTIONS OF PHYTOCHEMICALS WITH CELLULAR TARGETS

The phytochemical must be explicable in terms of chemical interactions with molecular targets in tissue. Although many phytochemicals produce effects in vitro in low concentration, there is no mystery in that it may produce obvious cellular response. One of the basic tenets of pharmacology is that drug molecules must exert chemical influence on one or multiple targets of cells in order to produce a pharmacological response. Likewise, a potent phytochemical should show pharmacological responses. Phytochemical compounds or drug molecules must bind to the constituent cellular molecules in such a way that the cellular function is altered. The chance of interaction with any cellular molecule would be negligible for nonspecific phytochemicals. Paul Ehrlich's principle remains true for the great majority of drug actions. Pharmacological effects can be observed with specific binding of drug molecules or phytochemicals with the cellular receptors [1].

These critical binding sites are often referred to as "drug molecular target." The mechanism of actions of a phytochemical with its target leads to a response that constitutes the thrust of pharmacological research into herbal medicine, which offers a rich source of novel drugs. Most drug targets are protein molecules that may form complex with lipids and metal clusters. Many anticancer drugs as well as mutagenic and carcinogenic agents interact specifically with DNA rather than protein.

PHYTOCHEMICAL BINDING WITH PROTEIN TARGETS

There are four main kinds of regulatory protein that are commonly involved as primary drug or phytochemical targets, namely,

1. Receptors
2. Enzymes
3. Transporter molecules
4. Ion channels

A few other types of protein are also known to function as drug targets. There exist many drug-producing sites of action that are not yet known. Many drugs are known to bind to multiple targets, including plasma proteins and a variety of cellular proteins without producing any physiological activity. Nevertheless, most drugs act on one or other of the aforementioned form types of proteins. Drug molecules would bind to the same cellular targets.

TARGETS FOR PHYTOCHEMICAL ACTION

A phytochemical entity is a chemical applied to a physiological system that affects its function in a specific way. Individual classes of phytochemicals bind only to certain targets and individual target recognize only certain classes of drug or phytochemical. No phytochemical is completely specific in their actions. In many cases, increasing the dose of a phytochemical will cause it to affect targets other than the principal target. This can lead to side effects.

PHYTOCHEMICAL–RECEPTOR INTERACTIONS

$$\text{Phytochemical A (agonist)} + R \underset{K_{-1}}{\overset{K_{+1}}{\rightleftharpoons}} AR \underset{\alpha}{\overset{\beta}{\rightleftharpoons}} AR^* \Longrightarrow \text{Response}$$

$$\text{Phytochemical B (antagonist)} + R \underset{K_{-1}}{\overset{K_{+1}}{\rightleftharpoons}} BR \Longrightarrow \text{No response}$$

The distinction between phytochemical binding and receptor activation is described earlier. The rate constants K_{+1}, K_{-1} and β and α apply to the binding and activation reactions, respectively. Phytochemical A is an agonist that leads to the activation of the receptor (R), whereas, phytochemical B is an antagonist. Binding occurs when more than one phytochemicals are present.

Phytochemicals act mainly on cellular targets, producing effects at biochemical, physiological, and structural levels. The direct effect of the phytochemical on its target produces acute responses, which may lead to downregulation of receptors, tolerance, or drug addiction. The cellular responses involve changes in gene expression associated with different functional levels.

In this section, we have avoided going into the details of the pharmacokinetics of herb–receptor interaction. The molecular details of how receptors work to produce their biological effects become more complicated in living organisms. Particular complications arise when it involves G-proteins in the reaction scheme. The consequences of receptor activation produce biological effects. Despite the shortcomings in herb–receptor interactions, the model proposed by Kenakin remains a useful theoretical basis for developing quantitative models of drug action [2]. A theoretical treatment model includes the knowledge of receptor interaction at the molecular level. It is important when considering phytochemicals in a therapeutic context. Their effects at different functional levels generally lead to secondary, delayed effects associated with efficacy or harmful effects of the phytochemicals. Phytochemicals that have immediate effects may take weeks to produce therapeutic benefits. Herbal medicine is often used to treat chronic conditions, and understanding long-term as well as acute herbal effects is becoming increasingly important. Western medicines have traditionally tended to focus on short-term physiological responses whereas herbal medicines traditionally tend to focus on long-term effects at the functional levels of a living organism.

SIGNALING PATHWAYS ASSOCIATED WITH p38 MAP KINASE

p38 MAPKs including α, β, γ, and δ are members of the MAPK that can be activated by a variety of environmental stress, oxidants, and phytochemicals. As with other MAPK cascades, the membrane component is a MAPKKK that phosphorylates and activates MKK 3/6, the p38 MAPK Kinases. MKK 3/6 can be activated directly by ASK1, which is initiated by apoptotic stimuli. p38 MAPK plays an important role in regulation of HSP27, MAPKAPK-2(MK2), and several transcription factors, including ATF-2, ElK-1, and CREB via activation of MSK-1.

G PROTEIN–COUPLED RECEPTORS SIGNALING

G protein-coupled receptors (GPCRs) are activated by phytochemicals. Upon receptor activation, the G protein exchanges guanosine diphosphate (GDP) for guanosine triphosphate (GTP), resulting in the dissociation of the GTP-bound α and β/γ subunits and subsequently triggering diverse signaling cascades. Receptors coupled to other G protein subtypes can lead to the activation of MAPK cascade. Src family Kinases are involved in the cascade activity, or signaling process through PyK2 and/or FAK. GPCRs can be activated by phytochemicals via PKC and CaMK11 resulting in the modulation of the MAPK pathway [3–5].

PROTEIN PHOSPHORYLATION AND KINASE CASCADE

One of the major principles to emerge from previous studies is that the protein phosphorylation is a key mechanism for controlling the function of proteins including enzymes, ion channels, transport protein, and protein receptors; this involves regulating cellular processes such as phosphorylation and dephosphorylation, which is accomplished by

1. Kinases
2. Phosphatases

Several hundred subtypes of these enzymes are represented in the human genome. Much effort is focused on mapping the complex interactions between signaling molecules and physiological processes such as inflammation.

The association of the two intracellular kinase domains allows autophosphorylation of intracellular tyrosine residues to take place. The phosphorylated tyrosine residues serve as docking sites for other intracellular proteins in signaling transduction cascade. The SH2 domain proteins were first identified in the Src oncogene product, which showed a highly conserved sequence of about 100 amino acids, forming a recognition site.

CONTROL OF RECEPTOR EXPRESSION

Receptor proteins are biosynthesized by the cells, and the level of expression is controlled via receptor-mediated pathways, which are themselves subjected to regulation [6–8]. Regulation of receptor function occurs through the following process:

1. Short-term regulation: desensitization
2. Long-term regulation: an increase or decrease of receptor expression

Examples of this type of control include the upregulation of various G-protein-coupled and cytokine receptors in response to inflammation. Long-term treatment with herbal medicines steadfastly induces adaptive responses that act on the functional levels. These are often the basis for therapeutic efficacy of herbal medicine.

The responses may take the form of a very slow onset of the pharmacological effects or the development of drug dependence, yet, the mechanism remains sketchy. It is most likely that changes in receptor expression are associated with functional properties of receptors and the type and concentrations of phytochemicals. The same principles apply to drug targets other than receptors such as ion channels, enzymes, and transporters, where changes in function follow long-term administration of herbal medicine that may result in resistance to certain anticancer agents.

RECEPTOR INTERACTIONS AND DISORDERS

Increasing understanding of receptor function in molecular terms has revealed the disease states linked to receptor interaction. The principal mechanisms of actions are the following:

1. Mutation in genes encoding receptors associated with signal transduction
2. Inhibition of receptor proteins by autoantibodies

Autoantibodies can mimic the effects of agonists as in thyroid hypersecretion, caused by activation of thyrotropin receptors. Activating antibodies have been reported in patients with severe hypertension and neurodegenerative disorder [9,10]. Inherited mutations of genes encoding G-protein-coupled receptors account for various diseases states. Receptor mutation can result in the activation of effector mechanisms in the absence of agonists. Many cancers are associated with mutations of the genes encoding growth factor receptors, kinase, and other proteins that play a critical role in signal transduction [11,12]. An effective anticancer phytochemical can have the ability to inhibit oncogene expression and suppress the signal transduction pathways [13].

TRANSCRIPTOME ANALYSIS IN COMBINATION WITH PATHWAY-BASED ACTIVITY

Transcriptome analysis in combination with signaling pathways is a helpful approach to understand the complex mechanism of actions of herbal constituents in cells. Genome-wide expression profiling showed activation of pathways and the affected biological processes. The identified molecular targets and pathways revealed several mechanisms that underlie the biological activity of the herbal extract [1]. Although extrapolation of genomic analysis is limited due to the sensitivity of the methodology, the study demonstrates the potential of the combination of focused strategies in analysis of multicomponent herbal mixture.

The link between a drug interacting with a molecular target and its effect at the functional level such as a change in serum glucose concentration or the shrinkage of a tumor, involves specific biological activity at the cellular level. Cells share the same repertoire of signaling pathways. The understanding of drug action at the cellular level is important. The regulation of cell function depends on the following components:

1. Free concentration of calcium ions
2. The storage and release of calcium ions by intracellular organelles
3. Calcium-dependent regulation of enzymes, including contractile proteins and vehicle proteins

As calcium concentration plays a vital role in cell function, a wide variety of herbal principle effects can result from interference with one or more of these mechanisms. Knowledge of the molecular and cellular activity has remarkably expanded in recent years. The therapeutic effects of herbal medicine can be better understood than before.

RELEASE OF CHEMICAL MEDIATORS IN CELLS

Much of pharmacology is based on disturbance with the body's own chemical mediators, particularly signaling molecules, hormones, and inflammatory molecules. The common mechanisms involved in the release of intrinsic chemical mediators are associated with the concentration of calcium. Drugs and phytochemicals that modulate the various control mechanisms also affect mediator release. Chemical mediators that are released from cells are classified as follows:

1. Mediators that are formed in storage vesicles and released by exocytosis, for example, hormones
2. Mediators that are produced and are released by diffusion, for example, nitric oxide

In the metabolism of drugs, calcium ions play a key role in the release of the chemical mediators. The concentration of calcium ions are affected by the action of drugs or phytochemicals.

RECEPTOR CONCEPT IS VERY IMPORTANT FOR DRUGS AND THERAPEUTIC AGENTS

Phytochemicals may mimic or antagonize the actions of hormones and neurotransmitters. Plant alkaloids such as atropine and receptor interaction provided a good model of the receptor concept, which had long been recognized as reflections of the interaction of plant alkaloids with their receptors and has led to the development of methods to develop new therapeutic agents targeted to specific receptors.

Hormone receptors are large proteins that contain active site(s) for interaction with drugs that bind them resulting in signal transduction. Hormone receptors have distinct domain to which plant alkaloids, both agonists and antagonists, bind. The link between the binding of a plant alkaloid and transduction may involve second messengers and a cascade of other proteins. In addition, other parts of a protein receptor can act as targets for different types of drugs or ligands.

FORMATION ACTIVE PHYTOCHEMICAL–RECEPTOR COMPLEX PRODUCES A CELLULAR RESPONSE

The active phytochemical–receptor complex initiates transduction at the cellular level. Many different classes of drugs or plant alkaloids may bind at a specific site resulting in conformational change in the receptor. If the ligand binds to a receptor without initiating a molecular response, the ligand is considered to be an antagonist.

In most cells the maximum cellular response to an agonist is achieved when only a small proportion of its receptors are occupied. Spare receptors increase the sensitivity of the cell according to changes in the agonist concentration.

An antagonist can inhibit the response to a second agonist by the activation of different receptors.

DOSE–RESPONSE CURVES TO MEASURE THE MECHANISMS OF DRUG ACTION

The quantitative analysis of phytochemical action is important in the development of new drugs, which should show the following properties:

1. Specificity in receptor interaction
2. Selectivity interaction with the receptor

The basic principle involves receptor interaction with phytochemicals:

1. When a phytochemical is present, phytochemical–receptor complexes are formed and dissociated in a dynamic equilibrium
2. Agonist–receptor complexes fluctuate between inactive and active conformations, resulting in the opening of ion channels or biochemical activity leading to a tissue response

3. The numbers of receptors expressed on a cell may change with the phytochemical concentration
4. By binding to receptors, antagonists reduce the number of receptors available for agonist

The principle represents how phytochemicals act in relaying empirical observations and experimental results to biochemical events. The quantitative measurements are manifested in the binding of radioactively labeled ligands (phytochemicals) to receptors isolated from tissues.

DOSE–RESPONSE RELATIONSHIP AND RECEPTOR INTERACTIONS

In drug development, it is important to categorize phytochemicals in terms of the biological systems with which they interact. The information can be used to characterize the molecular targets, whether they are agonists or antagonists. The approach is to measure the effects of phytochemicals on a variety of enzyme systems and tissue preparations that contain molecular targets. Thus, a phytochemical can be classified as an H_1 (histamine type 1 receptor) agonist if it acts in the same way as histamine and has its actions inhibited by an H_1 receptor antagonist in a competitive manner. An appropriate competitive antagonist may shift dose–response curves. If a chemical compound has effects that cannot be explained in terms of actions on known receptors, a new receptor may have been discovered. This can lead to a better understanding of how drugs or phytochemicals act, as well as the possible discovery of their respective receptors. Any biochemical responses to phytochemical effects such as accumulation of a second messenger may be an accurate measure of phytochemical action. It is essential when the phytochemical acts on molecular targets that are not present in tissue preparations. This type of biochemical response can provide a useful tool for the measurement of the phytochemical activity.

IDENTIFICATION OF NOVEL MOLECULAR TARGETS

Development in molecular biology has been instrumental in locating and identifying molecular targets, allowing them to be isolated, purified, and sequenced. This has facilitated the cloning of genes for the molecular target and has helped characterize how families of molecular targets are related. Identification of the DNA coding sequence for molecular targets enables us to obtain information on the mRNA carrying the message associated with the expression of the molecular target.

The findings have been very useful in identifying receptors that allow selective binding to phytochemicals. The growth of the bioinformatics and molecular biology has provided an important platform in our understanding of molecular targets and molecular mechanism of phytochemical actions.

ANTITUMOR EFFECTS OF DASATINIB

A novel drug named Dasatinib is a highly potent ATP-competitive Src/Abl kinase inhibitor with remarkable anticancer activity against solid tumors and chronic myeloid leukemia cell lines. Dasatinib has been in preclinical trial against various human cancers, including breast, prostate, pancreatic, lung, and head and neck. The findings showed Dasatinib inhibited the growth of laryngeal squamous cell carcinoma and Hep2 cells [14]. Dasatinib suppressed the expression of Bax, Bcl-2, caspase-3, and caspase-8 in the nude mouse xenograft model. The antitumor effect of Dasatinib was attributed to the induction of cell cycle arrest and consequently apoptosis via the suppression of Src phosphorylation. The results suggest a potential clinical application of Dasatinib for the treatment of various human cancers.

FENOFIBRATE AS AN ANTIPROLIFERATIVE DRUG

Fenofibrate, a peroxisome proliferator-androgen receptor-alpha agonist, is commonly used in treating hyperlipidemia and hypercholesterolemia. Recent studies have shown that fenofibrate exerts antiproliferative and pro-apoptotic activities in the prostate cancer cell line [15]. Fenofibrate induces cell cycle arrest in G1 phase and apoptosis in LNCaP cells. It reduces the gene expression of androgen receptor and the prostate-specific antigen and TMPRSS2, and it inhibited Akt phosphorylation. Fenofibrate can enhance the generation of intracellular reactive oxygen species and decrease superoxide dismutase activity in LNCaP cells. The therapeutic effects of fenofibrate are associated with preventing prostate cancer growth through inhibition of androgen activation and expression of the related genes in the signaling pathways.

NEW TARGET FOR AN OLD DRUG

The rational design of selective kinase inhibitors remains a subject for intensive research. Computer-aided design and modeling of the kinase inhibitor selectivity and the related data basis are available. The binding kinetics of 17 well-known kinase inhibitors against 143 kinases have been reported [16]. Based on the chemoinformatics, the binding profile of the chemotherapy drug mitoxantrone with five kinase targets was developed. Mitoxantrone was shown to exert low inhibitory properties against PIM1 kinase and could inhibit the PIM1-mediated phosphorylation in cancer cells. The x-ray crystal structure revealed a novel anticancer mode of PIM1 inhibition against kinase in contrary to the early study that showed mitoxantrone acted through DNA intercalation and type II topoisomerase inhibition.

ANTICANCER ACTIVITY OF TAXANES AND ANTHRACYCLINES

Although chemotherapy is commonly used for the treatment of cancer, chemotherapy medicines cause significant adverse side effects. A number of cancer drugs, given alone or in combination, are used as adjuvant therapy. This class of chemotherapy medicines includes taxanes and anthracyclines.

Taxanes chemotherapy medicines are Taxol, Taxotere, and Abraxane. Taxanes are mitotic inhibitors. They interfere with the ability of cancers to divide by inhibiting microtubule polymerization and can promote apoptosis [17,18]. Taxanes share similar mechanism of actions. Differences exist that may translate into differential efficacy among patients. The clinical success of the taxanes including paclitaxel and docetaxel in treating cancer patients has prompted a tremendous amount of research effort in the discovery of novel natural products and design of potent derivatives that possess the paclitaxel-like mechanism of action but with reduced side effects. Among the derivatives, the epothilones hold promise in combating paclitaxel-resistant tumors [19]. Future research in drug discovery can provide insights into the development of additional natural products with effective anticancer actions. New cancer drugs will provide treatment options for breast cancer resistant to Taxane and Anthracycline.

Anthracycline chemotherapy medicines are doxorubicin (Adriamycin), epirubicin (Ellence), daunorubicin (Cerubidine), mitoxantrone (Novantrone), and Daunorubicin. The first anthracycline was identified from *Streptomyces peucetius*. Subsequently, doxorubicin was developed. Shortly after, many other related compounds have been developed too. However, only a few are in clinical use. These compounds are used to treat many cancers, including lymphomas, leukemias, and breast, ovarian, uterine, and lung cancers. The anthracyclines are among the most effective chemotherapy medicines and are effective against more types of cancer than Taxanes [20].

MECHANISMS OF ACTIONS OF ANTHRACYCLINE

1. It inhibits both DNA and RNA synthesis by intercalating between the base pairs of the DNA or RNA strand, resulting in the inhibition of cancer cell growth [20].
2. It inhibits topoisomerase II enzyme and consequently blocks DNA transcription and replication. The binding of anthracycline with topoisomerase II mediates DNA-cleavage and prevents DNA repair by ligase [21].
3. It causes generation of iron-mediated oxygen radicals that induce DNA and protein damage. The free reactive oxygen species cause damage to cell membranes too [20].
4. It induces histone eviction from chromatin, resulting in interference in DNA damage response, transcriptome, and epigenome [22].

ADVERSE SIDE EFFECTS OF ANTHRACYCLINES

Anthracyclines are known to cause serious side effects such as vomiting, headache, and the worst of all, cardiotoxicity due to the inhibition of topoisomerase IIB in cardiomyocytes, with the formation of free radical in the heart. It also interferes with the ryanodine receptors of the sarcoplasmic reticulum. The metabolism of anthracyclines leads to bioaccumulation of toxic metabolites in the heart, and consequently, the cardiotoxicity leads to heart failure [23]. Preventive measures are needed to reduce the risk of cardiotoxicity by ways of controlling the cumulative doses (Table 2.1). Other measures include the use of cardioprotective agent such as dexrazoxane

TABLE 2.1
Four Levels of Phytochemical Actions

Mechanism of Action	Activity	Response
Molecular	Interaction with the molecular target	1. Activation of receptors
		2. Open or close of ion channels
		3. Activation of enzymes
		4. Transport protein
Cellular	Signal transduction	G protein-coupled receptor action
Tissue	Tissue functions	1. Proliferation
		2. Secretion
		3. Contraction
System	System functions	1. Cardiovascular
		2. Nervous

hydrochloride. Dexrazoxane is intended for protecting the heart against the cardiotoxic side effects of anthracyclines [24]. Like all chemotherapy medicines, taxanes and anthracyclines cause adverse side effects. The risk of heart problems can be more deleterious for cancer patients. Physicians are concerned with the higher risk of leukemia in women treated with doxorubicin. Many doctors preferred to use a combination of Herceptin with doxorubicin to treat HER-2 positive breast cancer. Unlike anthracyclines, taxanes are not known to cause cardiotoxicity. Therefore, a combination of chemotherapy medicines may offer better beneficial effects with fewer risks and side effects (see Table 2.1).

REFERENCES

1. Alexander, S.P., Mathie, A., and Peters, J.A. 2006. Guide to receptors and channels, 2nd edition. *Br. J. Pharmacol.* 147(Suppl. 3): S1–S168.
2. Kenakin, T. 2002. Drug efficacy at G protein-coupled receptors. *Annu. Rev. Pharmacol. Toxicol.* 42: 349–379.
3. Caunt, C.J., Finch, A.R., Sedgley, K.R., and McArdle, C.A. 2006. Seven-transmembrane receptor signalling and ERK compartmentalization. *Trends Endocrinol. Metab.* 17(7): 276–283.
4. Kim, E.K. and Choi, E.J. 2010. Pathological roles of MAPK signaling pathways in human diseases. *Biochim. Biophys. Acta* 1802(4): 396–405.
5. Aoki, Y., Niihori, T., Narumi, Y., Kure, S., and Matsubara, Y. 2008. The RAS/MAPK syndromes: Novel roles of the RAS pathway in human genetic disorders. *Hum. Mutat.* 29(8): 992–1006.
6. Donaldson, L.F., Hanley, M.R., and Villablanca, A.C. 1997. Inducible receptors. *Trends Pharmacol. Sci.* 18(5): 171–181.
7. Walaas, S.I. and Greengard, P. 1991. Protein phosphorylation and neuronal function. *Pharmacol. Rev.* 43(3): 299–349.
8. Alexander, S.P.H., Mathie, A., and Peters, J.A. 2004. Guide to receptors and channels, 1st edition. *Br. J. Pharmacol.* 141(Suppl. 1): S1–S126.
9. Spiegel, A.M. and Weinstein, L.S. 2004. Inherited diseases involving G proteins and G protein-coupled receptors. *Annu. Rev. Med.* 55: 27–39.
10. Cohen, P. 2002. Protein kinases—The major drug targets of the twenty-first century? *Nat. Rev. Drug. Discov.* 1(4): 309–315.
11. Klein, A., Wrulich, O.A., Jenny, M., Gruber, P., Becker, K., Fuchs, D., Gostner, JM., and Überall, F. 2013. Pathway-focused bioassays and transcriptome analysis contribute to a better activity monitoring of complex herbal remedies. *BMC Genom.* 14: 133.
12. Wermuth, C.G. 2004. Multitargeted drugs: The end of the "one-target-one-disease" philosophy? *Drug Discov. Today* 9(19): 826–827.
13. Gostner, J.M., Wrulich, O.A., Jenny, M., Fuchs, D., and Ueberall, F. 2012. An update on the strategies in multicomponent activity monitoring within the phytopharmaceutical field. *BMC Complement. Altern. Med.* 12: 18.
14. Song, Y., Sun, X., Bai, W.L., and Ji, W.Y. 2013. Antitumor effects of Dasatinib on laryngeal squamous cell carcinoma in vivo and in vitro. *Eur. Arch. Otorhinolaryngol.* 270: 1397–1404.

15. Zhao, H., Zhu, C., Qin, C., Tao, T., Li, J., Cheng, G., Li, P. et al. 2013. Fenofibrate down-regulates the expressions of androgen receptor (AR) and AR target genes and induces oxidative stress in the prostate cancer cell line LNCaP. *Biochem. Biophys. Res. Commun.* 432: 320–325.

16. Wan, X.B., Zhang, W., Li, L., Xie, Y.T., Li, W., and Huang, N. 2013. A new target for an old drug: Identifying mitoxantrone as a nanomolar inhibitor of PIM1 kinase via kinome-wide selectivity modeling. *J. Med. Chem.* 56: 2619–2629.

17. Kovar, J., Ehrlichova, M., Smejkalova, B., Zanardi, I., Ojima, I., and Gut, I. 2009. Comparison of cell death-inducing effect of novel taxane SB-T-1216 and paclitaxel in breast cancer cells. *Anticancer Res.* 29: 2951–2960.

18. Gallagher, Jr. B.M. 2007. Microtubule-stabilizing natural products as promising cancer therapeutics. *Curr. Med. Chem.* 14: 2959–2967.

19. Wood, A.J.J., Rowinsky, E.K., and Danehower, R.C. 1995. Paclitaxel (Taxol). *New Engl. J. Med.* 332: 1002.

20. Minotti, G., Menna, P., Salvatorelli, E., Cairo, G., and Gianni, L. 2004. Anthracyclines: Molecular advances and pharmacologic developments in antitumor activity and cardiotoxicity. *Pharmacol. Rev.* 56: 185–229.

21. Pommier, Y., Leo, E., Zhang, H., and Marchand, C. 2010. DNA topoisomerases and their poisoning by anticancer and antibacterial drugs. *Chem. Biol.* 17: 421–433.

22. Pang, B., Qiao, X., Janssen, L., Velds, A., Groothuis, T., Kerkhoven, R., Nieuwland, M. et al. 2013. Drug-induced histone eviction from open chromatin contributes to the thermotherapeutic effects of doxorubicin. *Nat. Commun.* 4: 1908.

23. Wang, H., Chan, Y.L., Li, T.-L., and Wu, C.-J. 2012. Improving cachectic symptoms and immune strength of tumor-bearing mice in chemotherapy by a combination of *Scutellaria baicalensis* and Wing-Shu-Yi-Qi-Tang. *Eur. J. Can.* 48(7): 1074–1084.

24. Lipshultz, S.E., Rifai, N., Daltan, V.M., Levy, D.E., Silverman, L.B., Lipsitz, S.R., Colan, S.D., and Asselin, B. 2004. The effect of dexrazoxane on myocardial injury in doxorubicin-treated children with acute lymphoblastic leukemia. *New Engl. J. Med.* 351: 145–153.

3 Combination of Cancer Drugs

The use of medical plants for treatment of ailments has played an important role in almost every culture on Earth. Herbal medicine was practiced by ancient cultures in Asia, Africa, Europe, and the Americas. The recent popularity in the use of herbs and plant products can be related to the ancient school of thought that herbs can provide some benefits over and above Western medicine and allow users to be personalized for individual needs. The widespread use of herbs has raised many scientific questions although Chinese medicine has been in practice for thousands of years. One of the major concerns is herbal safety. Herbs can interact with therapeutic drugs resulting in reducing or enhancing their efficacy. The interactions can pose some risk of inadvertently reducing the half-life of drugs such as indinavir, cyclosporine, and cyclophosphamide. In contrary, herbal products may act in a pathway similar to pharmaceuticals yet without undesirable side effects. Natural anti-inflammatory polyphenols and flavonoids are present in herbs and green tea too. As the use of nonsteroid anti-inflammatory drugs (NSAID) is associated with a reduced risk for cancers, it is probable that natural NSAID should be explored for cancer prevention [1]. The dosage of herbs that are aimed at the same function as the drugs such as sedative should be reduced to alleviate concerns about any effects that cause inhibition on the drug actions.

Chinese medicine is concerned with replenishing "qi"; thus, the modalities of Chinese medicine are used to compose the formula to provoke the "qi" before the pathological process manifests itself. While Chinese medicine has developed its own sophisticated repertoire of treatment for ailments including a variety of cancers, its primary emphasis is on restoring and preserving the healthy condition of the human body.

The hundreds of herbs that are commonly used in China and the West are rarely associated with adverse effects. The preponderance of evidence shows that when herbal medicine is used as an adjunct to Western medicine, Chinese herbs can enhance the pharmacological effects and attenuate the side effects.

Safety and herb–drug interactions have become complex with the increasing use of herbs by Westerners and are a legitimate concern for abuse and potential toxicity. Drug safety is always relative and determined by the health condition of a patient. Herb–drug interactions remain elusive.

Herbal medicines contain a variety of phytochemicals that may interact with each other, which ultimately can affect all body processes. The type of interactions determines the pharmacological or detrimental effects of herbal medicines. For example, Cinnabaris, a crude ore, contains mercuric oxide. Although it was considered unsafe by American Standards, it is still utilized in small doses in China for the short-term treatment of mental agitation without causing undesired effects. Many *Aristolochia* species have been shown to induce carcinogenic effects when continuously used for a period of 6 months or longer, yet these species continue to be used in China with positive results in the treatment of cancer.

In the United States, *Ephedra* has been made available over the counter as a natural energy and weight loss enhancer, yet it can cause high blood pressure, agitation, and insomnia. Abuse and misuse of *Ephedra* by professional health care providers have been reported. All in all, this has cast a serious doubt over the credibility and safety of Chinese herbs.

There is a paucity of data that describes the interactions between herbs and therapeutic drugs. A few herbs and foods show interactions with drugs especially used simultaneously. Tetracycline absorption can be impeded by milk-based foods, whereas grapefruit juice increases the blood volume of drugs such as antidepressants, antihistamines, and antihypertensive by inhibiting the drug metabolizing enzyme [2]. *Hypericum perforatum* is known to reduce blood levels of protease inhibitors by increasing their metabolism [3]. Green vegetables rich in vitamin K can antagonize the blood-thinning action of drugs such as Heparin or Warfarin [3]. *Gingko biloha* and *Salvia miltiorrhiza* promote microcirculation and inhibit platelet aggregation [4], yet they can potentiate the effects of anticoagulants, because of its immune-stimulating properties. *Glycyrrhiza uralensis* can mimic the action of cortisol to elevate blood pressure and increase fluid retention [4].

Previous study has found that there was a significant increase in taxane chemotherapy and decrease in anthracycline chemotherapy to treat breast cancer in the United States since 2005 [5]. Like all chemotherapy medicines, cancer drugs can cause serious health problems, including headache, vomiting, diarrhea, drowsiness, and heart problems. The health risk associated with cancer drugs can be higher when other treatments are used at the same time as a cancer drug. Many studies suggest that there is the growing trend of application of complementary and alternative medicine, especially herbal medicines in combination with conventional cancer therapy for cancer [6]. However, its efficacy remains to be investigated. It is believed that herbal medicines are used as an adjuvant therapy that can reduce toxicity of chemotherapy.

Compared to chemotherapy, Chinese herbal medicines in combination with chemotherapy drugs significantly improve the survival rate [6].

TABLE 3.1

Clinical Data Could Be Extracted from the Following Databases

Source	Period
EMBASE	1974–2012
MEDLINE	1946–2012
AMED	1985–2012
EBM	2005–2012
ACP	1991–2012
Database of Abstracts	2012
Cochrane Central Register	2012
Cochrane Methodology	2012
Health Technology Assessment	2012
China Knowledge Resource Integrated Database	1979–2012

This combined therapy significantly reduced the side effects related to cancer drugs and increased hemoglobin and platelet counts in cancer patients. The combined therapy remarkably increased tumor response to relative risk (RR) value of 1.36 with 95% confidence interval, 1.19–1.56, $p < 0.0005$ and improved Karnofsky performance score (KPS) (RR = 2.9, 95% CI = 1.62–5.18, $p = 0.0003$). This study demonstrated that Chinese herbal medicines as an adjuvant therapy can reduce toxicity related to cancer drugs, prolong survival rate, and improve KPS in advanced stage of cancer development.

Clinical trials were retrieved from 11 databases and published literature as shown in Table 3.1. The table shows the databases that are used to evaluate the effectiveness of chemotherapy in combination with herbal medicines.

Two types of KPS data—pre- and posttreatment—were reported in the studies. Significant findings with improvement were shown in the chemotherapy in combination with Chinese herbal medicines with the value of RR = 2.9, 95% CI = 1.62–5.18, $p = 0.0003$. The studies showed that the KPS of pretreatment had no obvious difference in combination therapy with Chinese herbal medicines while the posttreatment showed significant improvement in chemotherapy with Chinese herbal medicines compared to chemotherapy [7–9].

REDUCTION IN TOXICITY RELATED TO CHEMOTHERAPY

Adverse side effects including headache, nausea, and vomiting are the common undesirable effects that follow after chemotherapy. The combination chemotherapy with Chinese herbal medicines significantly attenuated toxicity grade III-IV compared to chemotherapy alone [7–9]. However, there was significant heterogeneity at toxicity grade I-IV probably due to difference in health conditions of cancer patients. Chemotherapy in combination with Chinese herbal medicine was shown to have different beneficial effects toward cancer patients. Recent studies showed that different combination strategies with selected herbal medicines have been recognized and used in cancer patients [10]. The combination therapy with herbal medicine as palliative care has become increasingly popular among cancer patients, yet the health benefits of the combined use of specific complementary and alternative medicines on different stages of cancer development remains to be further investigated. Different statistical tools such as meta-analyses are helpful to provide guidance to the evaluation of combination strategy with different herbal medicines. Most of the previous meta-analyses were set a good example with the inclusion criteria with Jadad score ≥3 to increase the reliability of the study [6]. Chemotherapy in combination with appropriate herbal medicines, especially those herbs with anticancer and anti-inflammatory properties, can significantly attenuate toxicity associated with chemotherapeutic drugs including nausea, vomiting, high blood pressure, headache, and organ toxicity. The health benefits of the combination strategy can improve the immune response of cancer patients with an increase in the peripheral blood leukocytes and hemoglobin. The combination therapy shows promise for those with advanced stage of non-small-cell lungs cancer patients and reduces the side effects of chemotherapy. The health benefits of combination therapy with herbal medicines as an adjuvant therapy were also reported in colorectal and nasopharyngeal cancer patients [6,11]. It is believed that herbal medicines are prescribed to patients to restore the balance of "Yin" and "Yang" of the human system and consequently alleviate the disease symptoms. The potential of integrating herbal medicines into chemotherapy remains to be evaluated in various cancers [12]. Different Chinese herbal medicines with anticancer and anti-inflammatory properties, including Radix Astragalus, Radix Ophiopogonis, and Radix Glycyrrhizae, show the tonifying qi, strengthening "yin," and cleaning blood stasis, the balance of which are important in cancer patients. A number of studies have reported that extracts of anticancer herbs can inhibit the growth of cancer cell lines and induce apoptosis. Radix Glycyrrhizae were shown to have anti-inflammatory and immunomodulation effects [13,14]. Other studies showed immunologic benefits of Radix Astragalus by activation of macrophage, and inhibiting T-helper cell type 2 cytokines [15]. The combination of herbal extracts of Radix Astragalus and Radix Angelicae increased white blood cells, hemoglobin, and platelet counts in the anemic rat model. The combination of Radix Astragulus and Angelicae enhanced erythropoietin production [16].

TABLE 3.2
Karnofsky Performance Score

Subject	Score	Complaint
Normal	100	No
Patient	70	Unable to conduct normal activity
Patient	50	Stationary with considerable assistance
Patient	40	Disable
Patient	30	Hospitalized

Overexpression of erythropoietin can reduce cisplatin toxicity in cells. However, the mechanism of action of Radix Astragalus remains sketchy. Nevertheless, the immune-stimulating effects and the attenuation of drug-induced toxicity were reported in lung cancer patients [17]. In addition, the improved survival rate, immediate tumor response, and alleviating suffering of cancer patients are attributed to the combination strategy with herbal medicines. It is expected that there are variations in efficacy of the combination strategy among cancer patients due probably to treatment duration and dietary factors. Individual health conditions and herbal extract concentration used in patients and more importantly the herb prescriptions can affect the pharmacological effects of herbal medicines on chemotherapy. Therefore, different biomarkers are to be measured to evaluate the efficacy of the combination treatment in cancer patients. The improved assessment method is based on the KPS shown in Table 3.2.

The clinical evaluation of health benefits based on clinical data and from the meta-analysis of the combination therapy with herbal medicine indicates that herbal medicine as adjuvant therapy has definite beneficial health effects on cancer patients. However, due to the complex drug–herbs interactions, only selected herbal medicines can be used to produce the pharmacological effects on cancer patients. A more defined integrated approach to chemotherapy with Chinese herbal medicine should be conducted using appropriate methodology of diagnosis and treatment for the evaluation of a combination therapy.

INTEGRATION OF *ASTRAGALUS* POLYSACCHARIDE WITH CISPLATIN

A platinum-based cancer drug regimen is the common curative method for lung cancer. However, platinum-based drug regimen causes adverse side effects; thus, the use of platinum-based drugs has its limitations. *Astragalus* polysaccharide isolated from the Radix of *Astragalus membranaceus* found its use in combination therapy with platinum-based drugs [18]. *Astragalus* polysaccharide could enhance tumor response and improve efficacy of chemotherapy drugs with reduction in toxicity related to cancer drugs. Lung cancer patients, after three cycles of treatment, showed significant differences in the overall quality of life ($p = 0.003$) with improved physical function and reduction in fatigue, pain, and nausea.

ANTICANCER PROPERTIES OF *CORNUS OFFICINALIS*

Selective hormone receptor inhibitors represent a common approach to therapeutic interventions for estrogen receptor-positive (ER+) breast cancer. The herbal medicine *Cornus officinalis* showed selective estrogen receptor modulation in combination with chemotherapeutic agents in the ER+ breast cancer cell line [19]. Treatment with these agents is associated with tumor resistance and undesirable harmful side effects. *C. officinalis* is commonly used in combination with therapeutic drugs for the treatment of breast cancer. Treatment of MCF-7 cells with *C. officinalis* caused an increase in the formation of the metabolite of α-hydroxyestrone, the metabolite of 17 beta-estradiol; a decrease in the formation of 16 alpha-hydroxyestrone, the pro-mitogenic metabolite; and a decrease in the formation of estrone. The molecular targets are related to modulation of estrogen receptor interactions. The findings suggested the molecular targets of herbal medicine in animal studies and implicated its clinical application to ER+ breast cancer.

SYNERGISTIC EFFECTS OF *TAXUS CUSPIDATA* WITH 5-FLUOROURACIL

The pharmacological effects of herbal medicine are increasingly recognized in various studies on cancer therapy. A combination of herbal medicines with cocktails of chemotherapeutics showed enhanced efficacy [20]. The *T. cuspidata* extract inhibited antitumor effects on a broad spectrum of cancer cell lines, including HL-60, BGC-823, KB, Bel-7402, and HeLa, but it showed only low inhibitory effects on normal mouse T/B lymphocytes, suggesting *T. cuspidate* causes low toxicity to normal cells. *T. cuspidate* induced apoptosis of cancer cells by inhibiting the G2/M phase of cell cycle. The *T. cuspidate* extract in combination with 5-fluorouracil inhibited cell growth of cancer cells with combination Index values ranging from 0.90 to 0.26 in MCF 7 cells [20]. The combination treatment of cancer cells with 5-fluorouracil exerts significant cytotoxicity in cancer cell lines. The results imply that the combination treatment with therapeutic agents may have palliative effects in cancer treatment.

PRECLINICAL TRIAL WITH PHY906 IN PATIENTS
WITH ADVANCED COLORECTAL CANCER

PHY906 is a novel Traditional Chinese medicine used to treat a broad spectrum of gastrointestinal disorders and related symptoms, including headache, fever, diarrhea, and abdominal pain, which are caused by therapeutic drugs [21]. The combination of irinotecan with PHY906 significantly improved antitumor activity and the survival rate in the animal studies. The combination of PHY906 with irinotecan, 5-fluorouracil, and leucovorin displayed additive antiproliferative effects on cancer cells. The clinical studies showed the safety and tolerability of PHY906, which did not affect the metabolism of 5-fluorouracil, irinotecan, and SN-38, the metabolite of irinotecan, which contributes to the pharmacological effects of PHY906.

FOOD SUPPLEMENTS AS ADJUVANT
OF CANCER THERAPY

Flavonoids are one of the major components of natural products, which are present in herbal medicines. Active flavonoids are taken in combination with anticancer drugs for the treatment of diseases and cancers. Some of the flavonoids can modulate P-glycoprotein (P-gp)-mediated anticancer drug disposition and cytotoxicity caused by cancer drugs [22]. Common flavonoids were tested for their biologic activity in vitro. Daunorubicin is chemotherapeutic of the anthracycline family used for the treatment of leukemia. Some of flavonoids decreased daunorubicin accumulation in vinblastine-resistant leukemia cells. In contrary, flavonoids that inhibited daunorubicin accumulation enhanced vinblastine accumulation, but did not affect colchicine accumulation. These flavonoids induced cytotoxicities by accumulation of cancer drugs and metabolites. Verapamil reversed the accumulation deficit and cytotoxicity of vinblastine, daunorubicin, and colchicine. The reversal of the accumulation is attributed to the biologic activity of flavonoids on P-gp-mediated ATPase activity, resulting in changes in cancer drug accumulation in drug-resistant cancer cells. The findings suggest that flavonoids may have different effects on chemotherapy drugs. This is important in designing combination strategies for cancer therapy.

EVIDENCE-BASED DRUG–HERBAL INTERACTIONS

The growing use of herbal medicines in combination with chemotherapy drugs has raised some concern about its health benefits and safety.

Health risks might arise from using herbal extracts in combination with cancer drugs. Studies showed interactions between natural products and cancer drugs are based on pharmacologic activity of natural products and metabolism of cancer drugs. Reports on clinical trials have indicated that interactions between herbal medicines and drugs may affect the efficacy of cancer drugs, and consequently, the beneficial effects of the combination therapy may be reduced [23]. However, detailed studies of the interaction between herbal medicines and drugs are lacking. More research on the integrated approach with herbal medicines is warranted.

ANTICANCER PROPERTIES OF THE TRADITIONAL CHINESE HERBAL FORMULA ZUO JIN WAN

An ancient Chinese herbal formula consisted of different herbs, known as Zuo Jin Wan (ZJW), was reported to exhibit anticancer properties [24]. The specific traditional Chinese medicine composition of ZJW showed that it could induce the reversal effects of multidrug resistance on colorectal cancer in both in vitro and in vivo studies. ZJW significantly enhanced the sensitivity of anticancer drugs in HCT116/L-OHP, SGC7901/DDP and Bel/Fu drug-resistant cells. A combination therapy with ZJW reversed the drug resistance of these cells, and inhibited the cancer cell growth in the colorectal tumor cells in a xenograft model. Inductive coupled plasma with mass spectrometry analysis showed that ZJW enhanced the accumulation of cancer drugs in HCT116/L-OHP cells in the mice model. It is believed that the reversal of drug resistance of colorectal cancer cells is related to the reduction of P-gp level in tumor. The anticancer properties of ZJW suggested that the specific herbal formulation can play an important role in reversing multidrug resistance of human cancers and that the pharmacological effects of herbs contribute to the modulation of activities of various systems, especially the immune system. The P-gp expression is considered as a useful molecular target for chemotherapy of different human cancers.

PHYTOCHEMICALS FOR THE HEART

The established formulations of herbal medicine have been used for the treatment of various ailments. These formulations are made up of different herbal ingredients. The specific formulations contain a variety of constituents including alkaloids, terpenoids, and saponin-like glycoside, etc. With a better understanding of the pharmacological activities of active principles, more drugs can be developed from the herbal medicines. Recent developments in herbal medicine for the prevention or treatment of cardiovascular disease have aroused a lot of research interest. The design of formulation is an integral approach to treat human diseases and involves processes of activity-directed composition and isolation of natural principal constituents. Studies showed increasing application of herbal formulations for cardiovascular disease [25–31]. The applicability of synergy-directed composition was demonstrated using *Hydrastis canadensis* [28]. Although individual herbs have pharmacological properties, a single herb may not show effective pharmacological activity. Herbal formulation is a common approach to the treatment of different diseases. The effectiveness of specific herbal formulations for cardiovascular disease is based on experience of Chinese medicine practitioners and the design of herbal composition to target treatment of cardiovascular symptoms [29].

PLANT ALKALOIDS

Naturally occurring plant alkaloids, in particular those identified from herbal medicines, are finding therapeutic use. Heart diseases can be well managed with specific formulations of herbal medicines. The combined action of multiple constituents of herbal medicines works with therapeutic benefits in humans [32]. The established formulations of traditional Chinese medicine show efficacy in the treatment of diseases. Some of the active alkaloids and terpenoids from medicinal herbs have been identified [32]. The pharmacological activities of these herbal compounds have been studied [32].

SAFFLOMIN

Safflower contains safflow yellow, carthamin, safflomin (Figure 3.1) and a lot other polyphenolic alkaloids, including pyrocatechol, dopa,

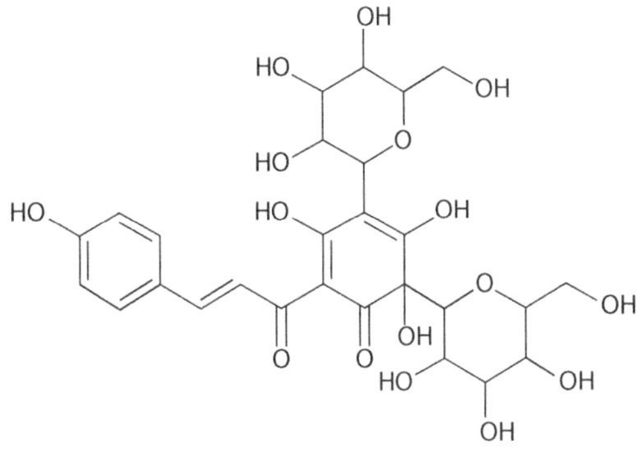

FIGURE 3.1   Chemical structure of safflomin.

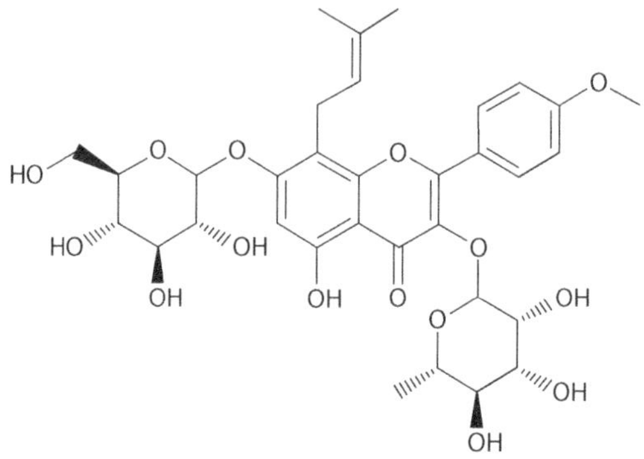

FIGURE 3.2 Chemical structure of icariin.

and kaempferol 3-rhamnoglucoside. Safflower contains a high proportion of polyunsaturated fats and is used as a protein supplement for livestock. Safflower is cultivated in China. Safflower is used in combination with Chinese angelica, zedoary turmeric rhizome (Rhizoma Zedoariae), and red-rooted salvia root (Radix Salviae Miltiorrhizae), to treat blood stasis with pain in the chest and abdomen and traumatic injuries due to obstruction of blood vessels [33,34]. Safflower can promote blood circulation by removing blood stasis, clear obstructions from blood vessels. Safflower is used for the treatment of coronary disease and Cerebral thrombosis. The dried flowers may be used to obtain carthamin (Figure 3.2), a red textile dye that was commercially important at one time but has since been replaced by synthetic aniline dyes, except in local areas of southwestern Asia. Safflower has been used as an adulterant of the condiment saffron.

ICARIIN

Icariin (Figure 3.2) from *Yinyanghuo* is a flavonol and PDE5 inhibitor. It induces the production of bioactive nitric oxide and acts like testosterone. It shows antioxidant and antidepressant effects in animal studies. Icariin was reported to have stimulating osteoblast activity [35]. The protective effects of icariin on vascular disease is attributed to its antioxidant properties. Phenolic compound such as *p*-hydroxybenzyl alcohol isolated from *Gastrodia elata blume* was reported to have pharmacological activity on ischemic brain injury through antioxidant activity related to gene expression. *p*-Hydroxybenzyl alcohol was shown to increase the transcriptional levels of isomerase and 1-Cys peroxiredoxin genes via reduction of oxidants [35]. *p*-Hydroxybenzyl alcohol is used to treat convulsive disorders [35].

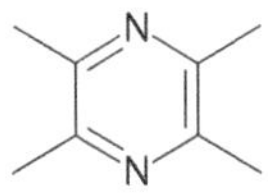

FIGURE 3.3   Chemical structure of ligustrazine (tetramethylpyrazine).

LIGUSTRAZINE

Ligustrazine (Figure 3.3), one of the active alkaloids from the rhizome of *Ligusticum walliichi*, has been reported to have pharmacological effects on a variety of vascular diseases [36]. The effects of ligustrazine on blood parameters, aorta and liver histology, and gene expression were investigated [36]. Ligustrazine decreased the total cholesterol, triglyceride, and low-density lipoprotein levels while it increased high-density lipoprotein level in the plasma. Ligustrazine could restore the total antioxidant capacity and superoxide dismutase 1 (SOD1) activity via inhibition of the induction of antioxidant genes both in aorta and in liver. Histological analysis revealed that ligustrazine suppressed atherosclerotic plaque progression in the thoracic aorta and lipid accumulation in the liver. Ligustrazine can alleviate oxidative stress and dyslipidemia probably due to its ability to remove superoxide and reduce nitric oxide production in human white blood cells of the immune system; thus, it strengthens the defense mechanism that is involved in defending the body against both infectious disease and foreign materials [37].

LEONURINE

Alkaloid leonurine was reported to show psychoactive properties. It is one of the alkaloids commonly found in the South African plant *Leonotis leonurus*, in the family Lamiaceae, and in the species *Leonotis nepetifolia* and *Leonotis artemisia*. An early study showed that leonurine (Figure 3.4) induced contraction in mouse uterine smooth muscle but relaxed in vascular smooth muscle of rat portal vein [38]. An in vitro study of leonurine showed that there were two types of smooth muscle

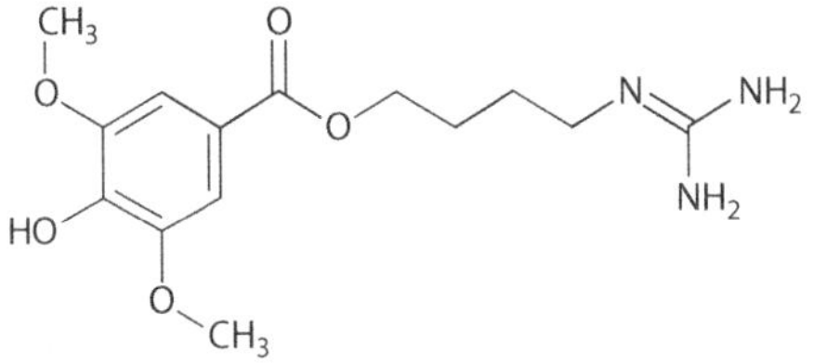

FIGURE 3.4   Chemical structure of leonurine.

associated with dual effects of leonurine on the contractile function of these smooth muscles. Stimulation and inhibition of the contraction in uterine and vascular smooth muscles were associated with the restoration of uterus after birth. The vasorelaxant effect could relieve pain and remove circulatory stagnation [38,39].

CORYDALIS

Corydalis belongs to a genus of about 300 species of plants native to north temperate areas and southern Africa. Most are weak-stemmed perennial garden plants with underground tubers and lobed or finely dissected leaves. Corydalis promotes blood circulation, activates the flow of qi, and kills pain. Corydalis tuber is an important Chinese herb that has been used to relieve almost any painful condition. It is used especially to relieve menstrual cramps and chest and abdominal pains. *Corydalis tuber* is used in combination with, for example, *Mongolian snakegourd* (Fructus Trichosanthis), onion bulb (*Bulbus Allii Macrostemi*), or red-rooted salvia root (Radix Salviae Miltiorrhizae). A variety of alkaloids have been separated from the rhizome of corydalis tuber, such as corydalis A, protopine (corydalis C), DL-tetrahydropalmatine (corydalis B), D-glaucine, coptisine, dehydrocorydaline, and corydalis G. Corydalis B has marked tranquilizing, hypnotic, and temperature-reducing effects. It is also a mild antiemetic. Various extracts from *C. tuber* have shown cardiotonic, hypotensive, and anticancer activity [40]. The cardiotonic properties may attribute to its antioxidant activity.

Vascular protection by naturally occurring polyphenols is believed to be associated with their antioxidant properties. Glycyrrhizin, one of the major active principles from licorice showed high potency in vitro and in vivo. Licorice is commonly used together with other herbs for the treatment of various diseases, including cardiovascular disease. Licorice has an array of pharmacological responses including inhibition of enzyme activity such as cyclooxygenase-1 (COX-1) and cyclooxygenase-2 (COX-2) [41–43]. Chemopreventive properties of licorice-active principles are believed to be related to its mediation of hepatic oxidative stress. Other studies showed that glycyrrhizin and glycyrrhetic acid, an active principle of licorice, inhibited growth of different cancer cell lines via apoptosis [44] and that tannins, an active principle of licorice, inhibited epidermal growth factor–induced cell transformations [45]. In addition, licorice extracts were reported to show effects on the expression of UDP-glucuronosyltransferases (UGTs) in rat hepatoma cells through transcriptional regulation of the genes associated with UGTs [46]. The overall pharmacology of licorice is complex.

The use of herbal medicine as an alternate to protect against cardiovascular diseases is the outgrowth of important findings of active principles and their specific pharmacological activity. Herbal components may be involved in mediation of antioxidative activity of enzyme such as dismutases, which are associated with cardiovascular diseases. COX-2 could also play an important role in antioxidative properties. COX-2 can convert arachidonic acid, a long-chained fatty acid to prostaglandins, which subsequently triggers inflammatory reactions in the body targeting at COX-2 [47].

ELLAGITANNINS

Ellagitannins are macromolecules formed when ellagic acid and a hydroxyl group of another molecule-like glucose bind together. It can be found in various fruits, including raspberries and blackberries. Different types of ellagitannins are reported to have various pharmacological activities, including antitumor effects [48]. It is believed that ellagitannins can lower the cholesterol level in blood; thus, it reduces incidence of heart disease. Ellagitannins can attenuate oxidative stress caused by free radicals.

TETRAMETHYLPYRAZINE

Tetramethylpyrazine is an active component of *L. wallichii Franchat*. It induces heme oxygenase-1 expression and attenuates myocardial ischemia/reperfusion injury in rats [49,50]. Superoxide anion production in ischemia myocardium was inhibited by tetramethylpyrazine. Tetramethylpyrazine was shown to suppress ischemia-induced ventricular arrhythmias and reduce the infarct size during ischemia/reperfusion injury in vivo. It is believed that tetramethylpyrazine displays antioxidant activity through the induction of heme oxygenase-1 and neutrophil inhibition [51]. The mode of action of tetramethylpyrazine is believed to be associated with its ability to scavenge superoxide anion and decrease nitric oxide production in human polymorphonuclear leukocytes. Other studies showed that tetramethylpyrazine has neuroprotective effects on the spinal cord in ischemia/reperfusion injury associated with the downregulation of spinal cord superoxide dismutase activity [52–54]. Also, an involvement of intracellular and extracellular Ca^{2+} in tetramethylpyrazine-induced colonic anion secretion was associated with its antioxidative activity. Tetramethylpyrazine inhibited Ca^{2+} ATPase of endoplasmic reticulum and promoted colonic mucosa secretion Cl^- and HCO_3^- via apical Cl^- channels and basolateral Na^+-K^+-Cl^- cotransporter and the basolateral diffusion of carbon dioxide [52].

An integral approach is promising for the treatment of human diseases including heart diseases with herbal medicine. Established formulations of herbal medicines have been used for the treatment of various disorders including cardiovascular diseases in China. The search for active principles from herbal medicines is warranted for the identification of candidates for drug development. However, its drug safety needs to be evaluated. It is important to understand the mechanism of the complex actions of herbal medicine when used for the treatment of cardiovascular diseases. Specific signal transduction pathways could be associated with the reduction of oxidative stress, and drug targets for the treatment of cardiovascular diseases. Herbs provide an excellent source for drug candidates.

The Chinese herb *Tripterygium wilfordii*, a vine-like plant found in South China, has been used as folk medicine. The water extract of *T. wilfordii* showed potency for the treatment of rheumatoid arthritis, nephritis, and systemic lupus erythematosus [55,56]. It was found to have antileukemia and antitumor properties [57,58]. Recent findings indicate that the active constituents from *T. wifordii*, especially the terpene family, contribute to anti-infective properties [59,60].

OLEANOLIC ACID

Oleanolic acid (Figure 3.5) is triterpene acid widely used as an anti-inflammatory agent. An earlier study showed the antibacterial activities of oleanolic acid against *Bacillus subtilis* and both methicillin-sensitive and methicillin-resistant *Staphylococcus aureus* [61]. Oleanolic acid also showed an inhibitory effect on the activation of Epstein-Barr virus early antigen (EBV-EA). The inhibitory activity is believed to be related to antitumor effects [62]. Another study reported that oleanolic acid methyl ester displayed weak antibacterial activity against *Micrococcus luteus* and *Escherichia coli* [63].

SALASPERMIC ACID

Salaspermic acid (Figure 3.6) is a tripterpene acid, with the molecular formula $C_{30}H_{48}O_4$ [12]. It was later isolated in *T. wifordii* [59,60]. It showed

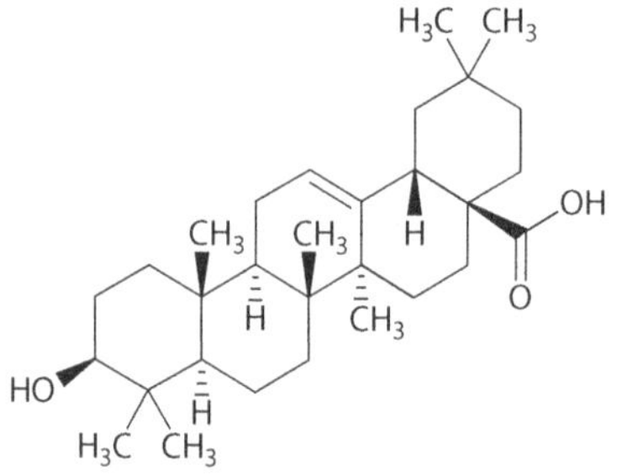

FIGURE 3.5   Chemical structure of oleanolic acid.

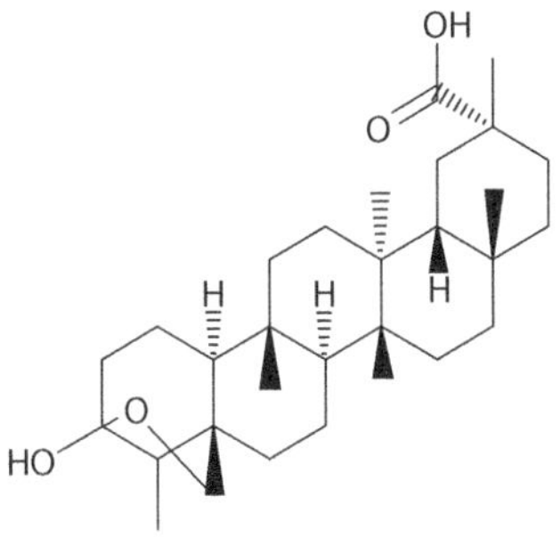

FIGURE 3.6 Chemical structure of salaspermic acid.

inhibition to HIV replication in H9 lymphocyte cells with an EC50 of 5 µg/mL. It was also found to have an inhibitory effect against HIV-1 reverse transcriptase activities. By comparing the anti-HIV activities of salaspermic acid and its derivatives, it was found that the acetal linkage in ring A and carboxyl group in ring B may be responsible for the anti-HIV activity [59,60].

TRIPTERYGIUM GENUS

Anti-HIV terpene compounds can also be found in the genus *Tripterygium*. Triptonine B (Figure 3.7), hypoglaunine B, and hyponine B were extracted from the root bark of *Tripterygium hypoglaucum* [64]. These compounds belonged to the class sesquiterpene alkaloids. "Sesquiterpene" refers to a terpene compound formed from three isoprene units. Among them,

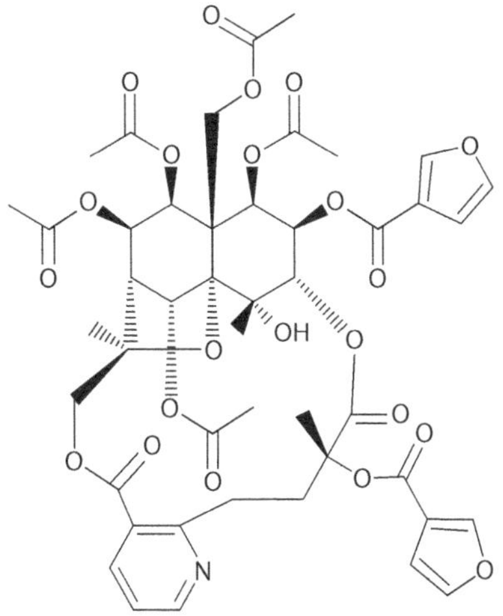

FIGURE 3.7 Chemical structure of triptonine B.

Triptonine B inhibited HIV replication in H9 lymphocytes. The in vitro therapeutic index (TI) of Triptonine B was found to be larger than 1000, while a TI>5 is considered significant in general.

FRIEDELIN AND DEHYDROABIETIC ACID

Friedelin (Figure 3.8) is another compound that had been isolated from the ethanol extract of *T. wilfordii* [65]. Friedelin is an active compound of the friedelane group of triterpenes. Previous studies indicated that friedelin exhibited antimicrobial activity against Gram-positive bacteria [66]. The tissue culture of *T. wilfordii* was investigated and the composition of compounds was found to be different from the plant itself, which included dehydroabietic acid and oleanolic acid [67]. Dehydroabietic acid is a kind of resin acid and is regarded as a potent antibacterial agent [68]. It was also found to have a synergistic antibacterial effect when combined with zinc oxide [69]. Compounds derived from dehydroabietic acid were also found to have antibacterial activities [70–72].

TERPENES

Terpenes and their derivatives were reported to show anti-infective properties. A series of new oxazines with potential anti-infectious and anti-parasitic activity were prepared from terpenes [73]. The design, synthesis, and biological evaluation of new oxazines with potential anti-infective activity were reported [74]. Tripterifordin, salaspermic acid, triptonine B, hypoglaunine B, and hyponine B were isolated from *T. wilfordii* and *Tripterygium* family from various studies [59,60,64,75]. These compounds belonged to the terpene family. The terpene family members are formed by biosynthesis from multiple isoprene units. Diterpene is a terpene compound formed from four isoprene units. The compound was an effective inhibitor of cytokine production [76]. Tripterifordin shows anti-HIV replication activities in H9 lymphocyte cells [59,60].

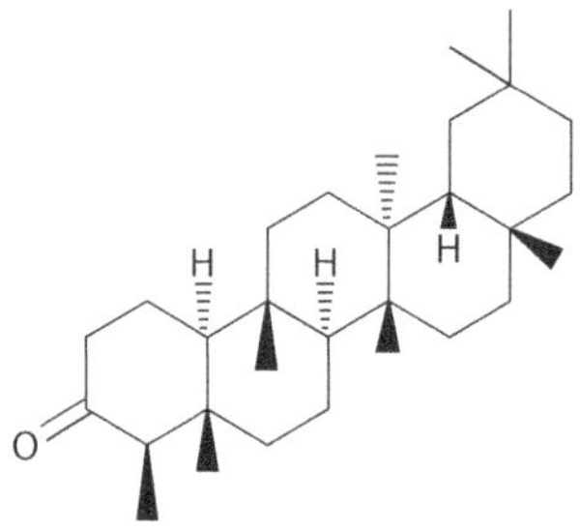

FIGURE 3.8   Chemical structure of friedelin.

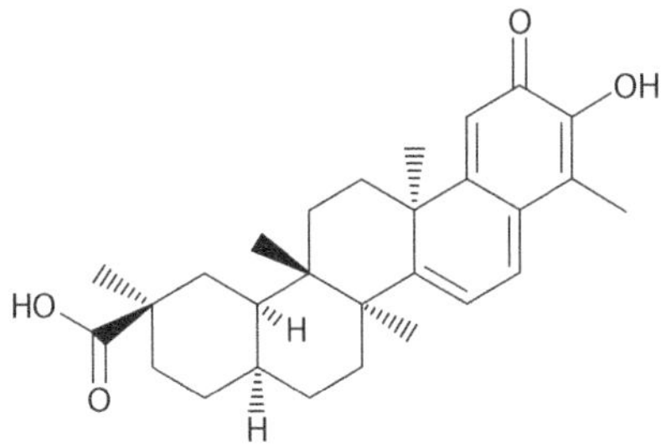

FIGURE 3.9 Chemical structure of celastrol.

Celastrol (Figure 3.9) is a quinone methide triterpene present in Celastraceae plants and is known to have a multitude of arrays of pharmacological activities. A common source of Celastrol is found in *T. wilfordii* Hook F, which is an ivy-like vine. Celastraceae has been used as a traditional medicine in China for hundreds of years. Celastrol has effectively been used in the treatment of autoimmune diseases, chronic inflammation, asthma, and neurodegenerative disease. Celastrol was shown to inhibit cancer cell proliferation and induce leukemic cell death in vitro. The medicinal properties of Celastrol and other terpenes are remarkable. Studies indicate that Celastrol shows different pharmacological activities associated with anti-infective properties [77].

COMBINATION OF PRINCIPLES OF HERBAL MEDICINE

Liquiritin, isoliquiritin, and isoliquirigenin are the active polyphenols present in *G. uralensis*, which has been used for the treatment of cancer and its complications (Figures 3.10 through 3.12). Recent study suggested that the combination of active phytochemicals show promise for lung cancer therapy with less toxicity [78,79]. *G. uralensis* has been shown to exhibit different pharmacological effects and antioxidant activity against oxidative stress, and is used for the treatment of various diseases including cancer [80–85]. Previous studies showed *G. uralensis* contained various types of triterpene, saponins, flavonoids, and phenolic compounds [79].

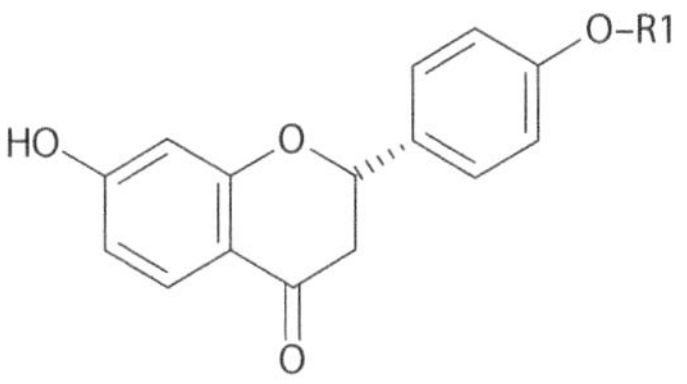

FIGURE 3.10 Chemical structure of liquiritin.

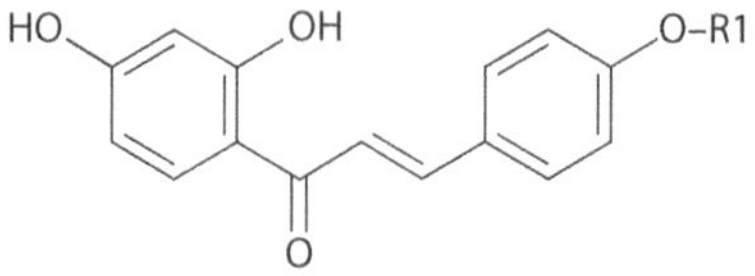

FIGURE 3.11   Chemical structure of isoquiritin.

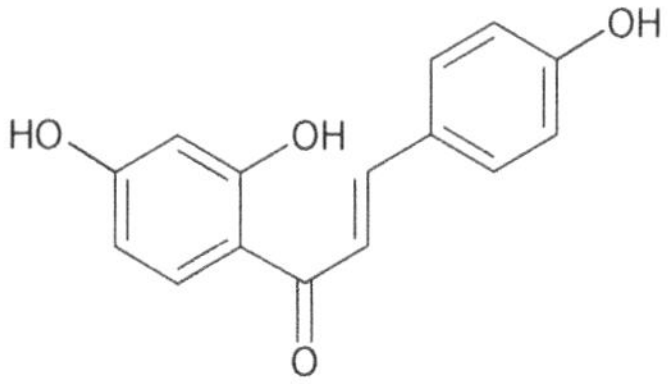

FIGURE 3.12   Chemical structure of isoliquirigenin.

However, the major ingredients of *G. uralensis* may have different biologic activities. Lactate dehydrogenase assays, such as FITC Annexin V staining assay, were performed to evaluate cellular cytotoxicity and apoptosis activity. The results showed that pretreatment with these polyphenols induced apoptosis in A549 cells. Liquiritin, isoquiritin, and isoliquirigenin significantly increased cytotoxicity of and upregulated p53 and p21 and downregulated the apoptotic pathways. Furthermore, it inhibited cell cycle at the G2/M phase. Western blot analysis showed it significantly decreased the protein expression of PCNA, MDM2, p-GSK3β, p-Akt, p-c-Raf, p-PTEN, caspase 3, pro-caspase 8, pro-caspase 9, PARP, and Bcl-2 in a concentration manner while the protein expression of p53, p21, and Bax was increased. In addition, the Akt pathway was downregulated. These findings suggest that liquiritin, isoquiritin, and isoliquirigenin inhibited p53-dependent pathway and showed crosstalk between Akt activities. These active polyphenols can be an alternative agent for the treatment of lung cancer.

The pharmacological activity of active fractions prepared from different methods can vary. Ethyl acetate extraction of *G. uralensis* appeared to show remarkable potency with antitumor activity in non-small-cell lung cancer while the aqueous extract showed little effects on non-small-cell lung cancer cell lines. The composition analysis of the ethyl acetate extract showed it contained three polyphenols: liquiritin, isoquiritin, and isoliquiritigenin. Although the individual principle present in *G. uralensis* showed anticancer effects on the lung cancer cell line, the combination of active principles prepared from *G. uralensis* exerts higher antitumor activities in lung cancer cells and in vivo study with the mice model.

MTT results showed that isoliquiritigenin exhibit the most potent anti-cancer activities. LDH assay and Annexin V staining confirmed that the combination of liquiritin, isoliquiritin, and isoliquiritigenin in the active fraction remarkably induced apoptosis [86].

The expression of related Bcl-2 family antiapoptotic protein Bcl-2, pro-apoptotic members Bax and Bid, and the cascade activity provide supportive evidence that the Bcl-2 and caspase cascades are involved in the apoptosis of A549 lung cancer cell line. The study revealed that the active extract containing liquiritin, isoliquiritin, and isoliquiritigenin upregulated Bax, Bid protein and downregulated Bcl-2. It increased the enzymatic activity of initiator caspase-8 and capsase-9. Subsequently it activated the effector caspase-3 and caspase-7, leading to the activation of intrinsic caspase cascades. Caspase-3 is known to be one of the key executioners of apoptosis because caspase-3 activation causes the cleavage or degradation of downstream PARP [87]. The results suggest the apoptosis of A549 cells is mediated through caspase-dependent activity.

The present findings showed that the expression level of Akt pathway is inhibited. Previous study reported that PI3K/Akt signaling cascade is related to cell survival and is mediated via the Bcl-2 family members or caspase family proteins [88]. Inhibition of Akt can downregulate phosphorylation of GSK-3β at Ser9 (inactive form of GSK-3β) to trigger apoptosis [89]. Akt is known to affect the activity of Erk pathway by silencing c-Raf via phosphorylation at the inhibitory site of Ser259 [90], implicating crosstalk among different signaling pathways. The deactivation of Akt by *G. uralensis* extract increased the activity of caspase-9 resulting in apoptosis; thus, blocking Akt can increase activities of GSK-3β, c-Raf, and caspase-9 in A549. Phosphorylated PTEN at Ser380 (suppressive residue) was blocked, suggesting the activation of PTEN through the mediation of the Akt pathway. The combination of polyphenols in *G. uralensis* caused inhibition of the expression level of phosphorylated PDK1 and Akt resulting in cell cycle arrest. Altered expression of regulatory cell cycle protein p53 and p21 was shown to promote G2/M phase arrest in A549 cells. Upregulation of p53 and p21 and downregulation of PCNA after treatment suggest it could regulate cell cycle arrest in A549 cancer cells. The findings suggest a crosstalk between p53, Bcl-2 family, caspase cascades, and Akt pathway [91,92].

Early studies reported that oxidative stress contributes to the pathogenesis of liver diseases. Hepato-toxicants and viral infections can induce liver damage and hepatitis [93,94]. The overproduction of reactive oxygen species from toxicant metabolism causes oxidative stress. The accumulation of the reactive oxygen species can induce liver injury [95]. A common approach to decrease the oxidative stress is by treatment with antioxidants [96]. *Ku Shen* is from the dried root of *Sophora flavescens* Ait, which has been commonly used for the treatment of liver diseases and hepatitis [97]. Oxymatrine (Figure 3.13) is one of the major alkaloids found in

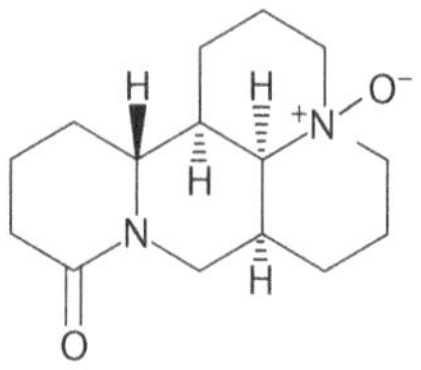

FIGURE 3.13   Chemical structure of oxymatrine.

Ku Shen. Previous studies reported the antitumor activities of *S. flaves-cens* Ait [98]. *S. flavescens* Ait has been used in herbal formulations to treat patients with liver cancer. It was reported that *S. flavescens* Ait could activate pregnane X receptor [99]. Among the principal alkaloids present in *S. flavescens* Ait, *N*-methylcytisine was identified as one of the active chemicals that can bind PXR through the induction of CYP3A4 genes. Other studies reported that oxymatrine exhibited more potent in vitro activity against human cancer cell lines, including HepG2 cells [98,100]. Oxymatrine killed the A375 cell line through the mediation of the mitochondrial membrane potential leading to the release of cytochrome c. It was believed that the low capacity of enzymes could be attributed to the reduced metabolism of oxymatrine in A375 cells. Although antitumor activities of oxymatrine have been reported, the mechanism of its action against cancer cells remains sketchy.

A previous study showed that oxymatrine exerted hepatoprotective activity against dimethyl sulfoxide toxicity in Clone 9 liver cells [101]. Pretreatment of cells with oxymatrine reduced cytotoxicity related to dimethyl sulfoxide. Oxymatrine exhibited scavenging effects on DMSO-induced oxidative stress in Clone 9 cells due probably to the inhibition of the metabolic activation of dimethyl sulfoxide. The results suggest that for cancer cells without oxymatrine pretreatment, the cell injury was implicated.

DMSO toxicity in Clone 9 cell line was increased by the incubation for a longer period of time. DMSO induced oxidative stress in Clone 9 cell line. Oxymatrine has the ability to reduce the cytotoxicity related to DMSO. Earlier studies suggested that DMSO induced the formation of reactive oxygen species [99,100,102]. The reduction of DMSO-induced cytotoxicity was associated with the antioxidative activity of oxymatrine; thus, oxymatrine showed protective effects on the cancer cell line. It appears that oxymatrine exhibited an inhibitory effect on reactive oxygen species formation in Clone 9 cell line, with radical scavenging activity against DMSO-induced cytotoxicity. Oxymatrine did not exhibit toxic effects on Clone 9 cells. The anticancer activity of oxymatrine on the cancer cell line was maintained after treatment with DMSO. A cell cycle analysis demonstrated that oxymatrine inhibited the cell proliferation of Clone 9 at M and G2 phases. The results provide useful information on the therapeutic potential of oxymatrine and its antioxidant activity in cancer cells.

REFERENCES

1. Wargovich, M.J., Woods, C., Hollis, D.M., and Zander, M.E. 2001. Herbals, cancer prevention and health. *J. Nutr.* 131(11 Suppl.): 3034S–3036S.
2. Hsu, H.Y. 1986. *Oriental Materia Medica*. Oriental Healing Arts Institute, Long Beach, CA, pp. 253–280.
3. Beinfield, H. and Korngold, E. 2003. Alternative therapies in health and medicine. *Chin. Med. Cancer Care* 9(5): 38–52.
4. Dharmananda, S. 2001. *The Interactions of Herbs and Drugs*. Institute for Traditional Medicine, Portland, OR.
5. Giordano, S.H., Lin, Y.L., Kuo, Y.F., Hortobagyi, G.N., and Goodwin, J.S. 2012. Decline in the use of anthracyclines for breast cancer. *J. Clin. Oncol.* 30: 2232–2239.
6. Li, S.G., Chen, H.Y., Ou-Yang, C.S., Wang, X.X., Yang, Z.J., Tong, Y., and Cho, W.C.S. 2013. The efficacy of Chinese herbal medicine as an adjunctive therapy for advanced non-small cell lung cancer: A systematic review and meta-analysis. *PLoS One* 8: e57604.
7. Huang, Y.L., Hou, A.J., Zhou, L., Hu, Y., and Shen, X.Y. 2012. Clinical study on non-small cell lung cancer treated by Zi Yin Qin Re herbs combined with chemotherapy. *Liaoning Tradit. Chin. Med. J.* 2: 53.
8. Huang, Y.L., Hou, A.J., Hu, Y., Zhou, L., and Gao, H.F. 2011. The clinical effect of Yi Qi Yang Yin Chinese medicine combined with GP chemotherapy to the advanced non-small cell lung cancer patients. *J. New Chin. Med.* 43: 47–49.
9. Zhu, L.M. and Guo, J.F. 2011. Clinical study of Yanshu injection on treating advanced non-small cell lung cancer. *Chin. J. Clin. Oncol. Rehabil.* 18: 380–384.
10. Chang, K.H., Brodie, R., Choong, M.A., Sweeney, K.J., and Kerin, M.J. 2011. Complementary and alternative medicine use in oncology: A questionnaire survey of patients and health care professionals. *BMC Cancer* 11: 196.
11. Cho, W.C. and Chen, H.Y. 2009. Clinical efficacy of traditional Chinese medicine as a concomitant therapy for nasopharyngeal carcinoma: A systematic review and meta-analysis. *Cancer Invest.* 27: 334–344.
12. Brake, M.K., Bartlett, C., Hart, R.D., Trites, J.R., and Taylor, S.M. 2011. Complementary and alternative medicine use in the thyroid patients of a head and neck practice. *Otolaryngol. Head Neck Surg.* 145: 208–212.
13. Roh, S.S., Kim, S.H., Lee, Y.C., and Seo, Y.B. 2008. Effects of radix adenophorase and cyclosporine A on an OVA-induced murine model of asthma by suppression to T cells activity, eosinophilia, and bronchial hyperresponsiveness. *Mediators Inflamm.* 2008: 781425.
14. Chan, H.T., Chan, C., and Ho, J.W. 2003. Inhibition of glycyrrhizic acid on aflatoxin B1-induced cytotoxicity in hepatoma cells. *Toxicology* 188: 211–217.

15. McCulloch, M., See, C., Shu, X.J., Broffman, M., Kramer, A., Fan, W.Y., Gao, J., Lieb, W., Shieh, K., and Colford, J.M. Jr. 2006. Astragalus-based Chinese herbs and platinum-based chemotherapy for advanced non-small-cell lung cancer: Meta-analysis of randomized trials. *J. Clin. Oncol.* 24: 419–430.

16. Chang, M.S., Kim, D.R., Ko, E.B., Choi, B.J., Park, S.Y., Kang, S.A., and Park, S.K. 2009. Treatment with Astragali radix and Angelicae radix enhances erythropoietin gene expression in the cyclophosphamide-induced anemic rat. *J. Med. Food* 12: 637–642.

17. Flower, A., Witt, L., Liu, J.P., Ulrich-Merzenich, G., and Yu, H. 2012. Guidelines for randomised controlled trials investigating Chinese herbal medicines. *J. Ethnopharmacol.* 140: 550–554.

18. Guo, L., Bai, S.P., Zhao, L., and Wang, X.H. 2012. *Astragalus* polysaccharide injection integrated with vinorelbine and cisplatin for patients with advanced non-small cell lung cancer: Effects on quality of life and survival. *Med. Oncol.* 29: 1656–1662.

19. Telang, N.T., Li, G., Sepkovic, D.W., Bradlow, H.L., and Wong, G.Y. 2012. Antiproliferative effects of Chinese herb *Cornus officinalis* in a cell culture model for estrogen receptor-positive clinical breast cancer. *Mol. Med. Rep.* 5: 22–28.

20. Shang, W., Qiao, J., Gu, C., Yin, W., Du, J., Wang, W., Zhu, M., Han, M., and Lu, W. 2011. Anticancer activity of an extract from needles and twigs of *Taxus cuspidate* and its synergistic effect as a cocktail with 5-fluorouracil. *BMC Complement. Altern. Med.* 11: 123.

21. Kummar, S., Copur, M.S., Rose, M., Wadler, S., Stephenson, J., O'Rourke, M., Brenckman, W. et al. 2011. A phase I study of the Chinese herbal medicine PHY906 as a modulator of irinotecan-based chemotherapy in patients with advanced colorectal cancer. *Clin. Colorectal Cancer* 10: 85–96.

22. Tran, V.H., Marks, D., Duke, R.K., Bebawy, M., Duke, C.C., and Roufogalis, B.D. 2011. Modulation of P-glycoprotein-mediated anticancer drug accumulation, cytotoxicity, and ATPase activity by flavonoid interactions. *Nutr. Cancer* 63: 435–443.

23. Chavez, M.L., Jordan, M.A., and Chavez, P.I. 2006. Evidence-based drug-herbal interactions. *Life Sci.* 78: 2146–2157.

24. Sui, H., Liu, X., Jin, B.H., Pan, S.F., Zhou, L.H., Yu, N.A., Wu, J. et al. 2013. Zuo Jin Wan, a traditional Chinese herbal formula, reverses P-gp-mediated MDR in vitro and in vivo. *Evid. Based Complement. Alternat. Med.* 2013: 957078.

25. Aljancic, I.S., Pesic, M., Milosavljevic, S.M., Todorovic, N.M., Jadranin, M., Milosavljevic, G., Povrenovic, D. et al. 2011. Isolation and biological evaluation of jatrophane diterpenoids from *Euphorbia dendroides*. *J. Nat. Prod.* 74(7): 1613–1620.

26. Guo, P., Li, Y., Xu, J., Liu, C., Ma, Y., and Guo, Y. 2011. Bioactive neo-clerodane diterpenoids from the whole plants of *Ajuga ciliata* Bunge. *J. Nat. Prod.* 74(7): 1575–1583.

27. Li, X.S., Zhou, X.J., Zhang, X.J., Su, J., Li, X.J., Yan, Y.M., Zheng, Y.T., Li, Y., Yang, L.M., and Cheng, Y.X. 2011. Sesquiterpene and norsesquiterpene derivatives from *Sanicula lamelligera* and their biological evaluation. *J. Nat. Prod.* 74(6): 1521–1525.

28. Junio, H.A., Sy-cordero, A.A., Ettefagh, K.A., Burns, J.T., Micko, K.T., Graf, T.N., Richter, S.J., Cannon, R.E., Oberlies, N.H., and Cech, N.B. 2011. Synergy-directed fractionation of botanical medicines: A case study with goldenseal (*Hydrastis canadensis*). *J. Nat. Prod.* 74(7): 1621–1629.

29. Arciniegas, A., Gonzalez, K., Perez-Castorena, K., Maldonado, J., Villasenor, J.L., and Romo de Vivar, A. 2011. Seco-eremophiladiolides and eremophilane glucosides from *Pittocaulon velatum*. *J. Nat. Prod.* 74(7): 1584–1589.

30. Zhu, Y.Z., Huang, S.H., Tan, B.K.H., Sun, J., Whiteman, M., and Zhu, Y.C. 2004. Antioxidants in Chinese herbal medicines: A biochemical perspective. *Nat. Prod. Rep.* 21(4): 478–489.

31. Afifi-Yazar, F., Kasabri, V., and Abu-Dahab, R. 2011. Medicinal plants from Jordan in the treatment of diabetes: Traditional uses vs. in vitro and in vivo evaluations—Part 2. *Planta Med.* 77(11): 1210–1220.

32. Ho, J.W., Cheung, M.W., and Yu, V.W. 2012. Active phytochemicals from Chinese herbs as therapeutic agents for the heart. *Cardiovasc. Hematol. Agents Med. Chem.* 10(3): 251–255.

33. Wang, C.C., Choy, C.S., Liu, Y.H., Cheah, K.P., Li, J.S., Wang, J.T.J., Yu, W.Y., Lin, C.W., Cheng, H.W., and Hu, C.M. 2011. Protective effect of dried safflower petal aqueous extract and its main constituent, carthamus yellow, against lipopolysaccharide-induced inflammation in RAW264.7 macrophages. *J. Sci. Food Agric.* 91(2): 218–225.

34. Meselhy, M.R., Kadota, S., Momose, Y., Hatakeyama, N., Kusai, A., Hattori, M., and Namba, T. 1993. Two new quinochalcone yellow pigments from *Carthamus tinctorius* and Ca^{2+} antagonistic activity of tinctormine. *Chem. Pharm. Bull.* (Tokyo) 41(10): 1796–1802.

35. Ning, H., Xin, Z.C., Lin, G., Banie, L., Lue, T.F., and Lin, C.S. 2006. Effects of icariin on phosphodiesterase-5 activity in vitro and cyclic guanosine monophosphate level in cavernous smooth muscle cells. *Urology* 68(6): 1350–1354.

36. Jiang, F., Qian, J., Chen, S., Zhang, W., and Liu, C. 2011. Ligustrazine improves atherosclerosis in rat via attenuation of oxidative stress. *Pharm. Biol.* 49(8): 856–863.

37. Liao, S.L., Kao, T.K., Chen, W.Y., Lin, Y.S., Chen, S.Y., Raung, S.L., Wu, C.W., Lu, H.C., and Chen, C.J. 2004. Tetramethylpyrazine reduces ischemic brain injury in rats. *Neurosci. Lett.* 372(1–2): 40–45.

38. Liu, X.H., Pan, L.L., Chen, P.F., and Zhu, Y.Z. 2010. Leonurine improves ischemia-induced myocardial injury through antioxidative activity. *Phytomedicine* (Jena) 17(10): 753–759.

39. Liu, X., Pan, Li, L., Gong, Q., and Zhu, Y. 2010. Leonurine (SCM-198) improves cardiac recovery in rat during chronic infarction. *Eur. J. Pharmacol.* 649(1–3): 236–241.

40. Wu, L., Ling, H., Li, L., Jiang, L., and He, M. 2007. Beneficial effects of the extract from *Corydalis yanhusuo* in rats with heart failure following myocardial infarction. *J. Pharm. Pharmacol.* 59(5): 695–701.

41. Rahman, S. and Sultana, S. 2006. Chemopreventive activity of glycyrrhizin on lead acetate mediated hepatic oxidative stress and its hyperproliferative activity in Wistar rats. *Chem. Biol. Interact.* 160(1): 61–71.

42. Yehuda, I., Madar, Z., Szuchman-Sapir, A., Tamir, S., and Glabridin, A. 2011. A phytoestrogen from licorice root, up-regulates manganese superoxide dismutase, catalase and paraoxonase 2 under glucose stress. *Phytother. Res.* 25(5): 659–667.
43. Chandrasekaran, C.V., Deepak, H.B., Thiyagarajan, P., Kathiresan, S., Sangli, G., Deepak, M., and Agarwal, A. 2011. Dual inhibitory effect of *Glycyrrhiza glabra* (GutGard (TM)) on COX and LOX products. *Phytomedicine* (Jena) 18(4): 278–284.
44. Hibasami, H., Iwase, H., Yoshioka, K., and Takahashi, H. 2006. Glycyrrhetic acid (a metabolic substance and aglycon of glycyrrhizin) induces apoptosis in human hepatoma, promyelotic leukemia and stomach cancer cells. *Int. J. Mol. Med.* 17(2): 215–219.
45. Nomura, M., Tsukada, H., Ichimatsu, D., Ito, H., Yoshida, T., and Miyamoto, K.I. 2005. Inhibition of epidermal growth factor-induced cell transformation by tannins. *Phytochemistry* 66(17): 2038–2046.
46. Leung, Y.K. and Ho, J.W. 2000. Effects of licorice on the expression of UGT1A isozymes in rat hepatoma cells. *FASEB J.* 14(8): A1525.
47. Ni, Y., Kuai, J., Lu, Z., Yang, G., Fu, H., Wang, J., Tian, F., Yan, X., Zhao, Y., Wang, Y., and Jiang, T. 2011. Glycyrrhizin treatment is associated with attenuation of lipopolysaccharide-induced acute lung injury by inhibiting cyclooxygenase-2 and inducible nitric oxide synthase expression. *J. Surg. Res.* 165(1): E29–E35.
48. Ito, H. 2011. Metabolites of the ellagitannin geraniin and their antioxidant activities. *Planta Med.* 77(11): 1110–1115.
49. Cheng, C.Y., Sue, Y.M., Chen, C.H., Hou, C.C., Chan, P., Chu, Y.L., Chen, T.H., and Hsu, Y.H. 2006. Tetramethylpyrazine attenuates adriamycin-induced apoptotic injury in rat renal tubular cells NRK-52E. *Planta Med.* 72(10): 888–893.
50. Pramanik, S.S., Sur, T.K., Debnath, P.K., Pramanik, T., and Bhattacharyya, D. 2011. Effect of *Pueraria tuberosa* on cold immobilization stress induced changes in plasma corticosterone and brain monoamines in rats. *J. Nat. Remedies* 11: 69–75.
51. Hsiao, G., Chen, Y.C., Lin, J.H., Lin, K.H., Chou, D.S., Lin, C.H., and Sheu, J.R. 2006. Inhibitory mechanisms of tetramethylpyrazine in middle cerebral artery occlusion (MCAO)-induced focal cerebral ischemia in rats. *Planta Med.* 72(5): 411–417.
52. Zhu, J.X., Zhang, G.H., Yang, N., Wong, H.Y.C., Chung, Y.W., and Chan, H.C. 2006. Involvement of intracellular and extracellular Ca^{2+} in tetramethylpyrazine-induced colonic anion secretion. *Cell Biol. Int.* 30(6): 547–552.
53. Zhao, W.C., Duan, D.X., Wang, Z.J., Tang, N., Yan, M., Zhang, G.H., and Xing, Y. 2005. The underlying cellular mechanism in the effect of tetramethylpyrazine on the anion secretion of colonic mucosa. *Jpn. J. Physiol.* 55(6): 325–329.
54. Giner, E., El Alami, M., Manez, S., Recio, M.C., Rios, J.L., and Giner, R.M. 2011. Phenolic substances from *Phagnalon rupestre* protect against 2,4,6-trinitrochlorobenzene-induced contact hypersensitivity. *J. Nat. Prod.* 74(5): 1079–1084.

55. Tao, X., Younger, J., Fan, F.Z., Wang, B., and Lipsky, P.E. 2002. Benefit of an extract of *Tripterygium wilfordii* Hook F in patients with rheumatoid arthritis: A double-blind, placebo-controlled study. *Arthritis Rheum.* 46(7): 1735–1743.

56. Ho, L.J. and Lai, J.H. 2004.Chinese herbs as immunomodulators and potential disease-modifying antirheumatic drugs in autoimmune disorders. *Curr. Drug. Metab.* 5: 181–192.

57. Kupchan, S.M. and Schubert, R.M. 1974. Selective alkylation: A biomimetic reaction of the antileukemic triptolides? *Science* 185: 791–793.

58. Yang, S., Chen, J., Guo, Z., Xu, X.M., Wang, L., Pei, X.F., Yang, J., Underhill, C.B., and Zhang, L. 2003. Triptolide inhibits the growth and metastasis of solid tumors. *Mol. Cancer Ther.* 2: 65–72.

59. Chen, K., Shi, Q., Kashiwada, Y., Zhang, D.C., Hu, C.Q., Jin, J.Q., Nozaki, H., Kilkuskie, R.E., Tramontano, E., and Cheng, Y.C. 1992. Anti-aids agents, 6. Salaspermic acid, an anti-HIV principle from *Tripterygium wilfordii*, and the structure-activity correlation with its related compounds. *J. Nat. Prod.* 55(3): 340.

60. Chen, K., Shi, Q.A., Fujioka, T., Zhang, D.C., Hu, C.Q., Jin, J.Q., Kilkuskie, R.E., and Lee, K.H. 1992. Anti-AIDS agents, 4. Tripterifordin, a novel anti-HIV principle from *Tripterygium wilfordii*: Isolation and structural elucidation. *J. Nat. Prod.* 55(1): 88–92.

61. Khera, S., Woldemichael, G. M., Singh, M. P., Suarez, E., and Timmermann, B.N. 2003. A novel antibacterial iridoid and triterpene from *Caiophora coronata. J. Nat. Prod.* 66(12): 1628.

62. Taniguchi, S., Imayoshi, Y., Kobayashi, E., Takamatsu, Y., Ito, H., Hatano, T., Sakagami, H. et al. 2002. Production of bioactive triterpenes by *Eriobotrya japonica* calli. *Phytochemistry* 59: 315–323.

63. Weimann, C., Göransson, U., Pongprayoon-Claeson, U., Claeson, P., Bohlin, L., Rimpler, H., and Heinrich, M. 2002. Spasmolytic effects of baccharis conferta and some of its constituents. *J. Pharm. Pharmacol.* 54(1): 99–104.

64. Duan, H., Takaishi, Y., Imakura, Y., Jia, Y., Li, D., Cosentino, L.M., and Lee, K.H. 2000. Sesquiterpene alkaloids from *Tripterygium hypoglaucum* and *Tripterygium wilfordii*: A new class of potent anti-HIV agents. *J. Nat. Prod.* 63(3): 357–361.

65. Yang, J.H., Luo, S.D., Wang, Y.S., Zhao, J.F., Zhang, H.B., and Li, L. 2006. Triterpenes from *Tripterygium wilfordii* Hook. *J. Asian Nat. Prod. Res.* 8(5): 425–429.

66. Pretto, J.B., Cechinel-Filho, V., Noldin, V.F., Sartori, M.R., Isaias, D.E., and Cruz, A.B. 2004. Antimicrobial activity of fractions and compounds from *Calophyllum brasiliense* (Clusiaceae/Guttiferae). *Z. Naturforsch. C* 59: 657–662.

67. Kutney, J.P., Hewitt, G.M., Kurihara, T., Salisbury, P.J., Sindelar, R.D., Stuart, K.L., Townsley, P.M., Chalmers, W.T., and Jacoli, G.G. 1981. Cytotoxic diterpenes triptolide, tripdiolide, and cytotoxic triterpenes from tissue cultures of *Tripterygium wilfordii*. *Can. J. Chem.* 59: 2677.

68. Soderberg, T.A., Gref, R., Holm, S., Elmros, T., and Hallmans, G. 1990. Antibacterial activity of rosin and resin acids in vitro. *Scand. J. Plast. Reconstr. Surg. Hand Surg.* 24: 199–205.

69. Soderberg, T.A., Holm, S., Gref, R., and Hallmans, G. 1991. Antibacterial effects of zinc oxide, rosin, and resin acids with special reference to their interactions. *Scand. J. Plast. Reconstr. Surg. Hand Surg.* 25(1): 19.

70. Tapia, A.A., Vallejo, M.D., Gouiric, S.C., Feresin, G.E., Rossomando, P.C., and Bustos, D.A. 1997. Hydroxylation of dehydroabietic acid by *Fusarium* species. *Phytochemistry* 46(1): 131.

71. Savluchinske, F.S., Gigante, B., Roseiro, J.C., and Marcelo-Curto, M.J. 1999. Antimicrobial activity of diterpene resin acid derivatives. *J. Microbiol. Methods* 35(3): 201.

72. Savluchinske-Feio, S., Gigante, B., and Roseiro, J.C. 2006. Antimicrobial activity of resin acid derivatives. *Appl. Microbiol. Biotechnol.* 72(3): 430.

73. Posner, G.H. and O'Neill, P.M. 2004. Knowledge of the proposed chemical mechanism of action and cytochrome p450 metabolism of antimalarial trioxanes like artemisinin allows rational design of new antimalarial peroxides. *Acc. Chem Res.* 37(6): 397.

74. Sindhu, T.J., Sonia, D.A., Girly, V., Meena, C., Bhat, A.R., and Krishakumar, K. 2013. Biological activities of oxazine and its derivatives: A review. *Int. J. Pharma. Sci. Res.* 4(11): 134–143.

75. Lei, W. and Li, X.Y. 1991. Immune suppressive actions of celastrol, a triterpene compound from *Tripterygium wilfordii*. *Pharmacol. Clin. Chin. Mat. Med.* 7: 18.

76. Duan, H., Takaishi, Y., Momota, H., Ohmoto, Y., Taki, T., Jia, Y., and Li, D. 2001. Immunosuppressive sesquiterpene alkaloids from *Tripterygium wilfordii*. *J. Nat. Prod.* 64: 582–587.

77. Lee, N.H. and Ho, J.W. 2008. Celastrol and terpenes as anti-infective agents. *Antiinfect. Agents Med. Chem.* 7(2): 97–100.

78. Jeong, S.J., Koh, W., Kim, B., and Kim, S.H. 2011. Are there new therapeutic options for treating lung cancer based on herbal medicines and their metabolites? *J. Ethnopharmacol.* 138: 652–661.

79. Zhang, Q.Y. and Ye, M. 2009. Chemical analysis of the Chinese herbal medicine Gan-Cao (licorice). *J. Chromatogr. A* 1216: 1954–1969.

80. Yo, Y.T., Shieh, G.S., Hsu, K.F., Wu, C.L., and Shiau, A.L. 2009. Licorice and Licochalcone-A induce autophagy in LNCaP prostate cancer cells by suppression of Bcl-2 expression and the mTOR pathway. *J. Agric. Food Chem.* 57: 8266–8273.

81. Dong, S.J., Inoue, A., Zhu, Y., Tanji, M., and Kiyama, R. 2007. Activation of rapid signally pathways and the subsequent transcriptional regulation for the proliferation of breast cancer MCF-7 cells by the treatment with an extract of *Glycyrrhiza glabra* root. *Food Chem. Toxicol.* 45: 2470–2478.

82. Fu, Y., Hsieh, Z.C., Guo, J.Q., Kunicki, J., Lee, Y.W.T., Darzynkiewicz, Z., and Wu, J.M. 2004. Licochalcone-A, a novel flavonoid isolated from licorice root (*Glycyrrhiza glabra*), causes G2 and late-G1 arrests in androgen-independent PC-3 prostate cancer cells. *Biochem. Biophys. Res. Commun.* 322: 263–270.

83. Hsiang, C.Y., Lai, I.L., Chao, D.C., and Ho, T.Y. 2002. Differential regulation of activator protein 1 activity by glycyrrhizin. *Life Sci.* 70: 1643–1656.

84. Satomi, Y., Nishino, H., and Shibata, S. 2005. Glycyrrhetinic acid and related compounds induce G1 arrest and apoptosis in human hepatocellular carcinoma HepG2. *Anticancer Res.* 25: 4043–4047.

85. Yoon, G., Bok, Y.K., and Seung, H.C. 2007. Topoisomerase I inhibition and cytotoxicity of licochalcones A and E from *Glycyrrhiza inflate*. *Arch. Pharm. Res.* 30: 313–316.

86. Zhou, Y. and Ho, W.S. 2014. Combination of liquiritin, isoliquiritin and isoliquirigenin induce apoptotic cell death through upregulating p53 and p21 in the A549 non-small cell lung cancer cells. *Oncol. Rep.* 31(1): 298–304.

87. Chang, H.Y. and Yang, X. 2000. Proteases for cell suicide: Functions and regulation of caspases. *Microbiol. Mol. Biol. Rev.* 64: 821–846.

88. Hsarkar, F. and Li, Y.W. 2007. Targeting multiple signal pathways by chemopreventive agents for cancer prevention and therapy. *Acta Pharmacol. Sin.* 28: 1305–1315.

89. Beurel, E. and Jope, R.S. 2007. The paradoxical pro- and anti-apoptotic actions of GSK3 in the intrinsic and extrinsic apoptosis signaling pathways. *Prog. Neurobiol.* 79: 173–189.

90. Rommel, C., Clarke, B.A., Zimmermann, S., Nuñez, L., Rossman, R., Reid, K., Moelling, K., Yancopoulos, G.D., and Glass, D.J. 1999. Differentiation stage-specific inhibition of the Raf-MEK-ERK pathway by Akt. *Science* 286: 1738–1741.

91. Gottlieb, T.M., Martinez-Leal, J.F., Seger, R., Taya, Y., and Oren, M. 2002. Cross-talk between Akt, p53 and MDM2: Possible implications for the regulation of apoptosis. *Oncogene* 21: 1299–1303.

92. Mayo, L.D. and Donner, D.B. 2002. The PTEN, MDM2, p53 tumor suppressor-oncoprotein network. *Trends Biochem. Sci.* 27: 462–467.

93. Chui, C.H., Lau, F.Y., Tang, J.C.O., Kan, K.L., Cheng, G.Y.M., Wong, R.S.M., Kok, S.H.L., Lai, P.B.S., and Ho, R. 2005. Gambari R and Chan ASC: Activities of fresh juice of *Scutellaria barbata* and warmed water extract of Radix Sophorae Tonkinensis on anti-proliferation and apoptosis of human cancer cell lines. *Int. J. Mol. Med.* 16(2): 337–341.

94. Selvaraj, V., Yeager-Armstead, M., and Mindy, E. 2012. Protective and antioxidant role of selenium on arsenic trioxide-induces oxidative stress and genotoxicity in the fish hepatoma cell line PLHC-1. *Environ. Toxicol. Chem.* 31(12): 2861–2869.

95. Brito, N.J.N., Lopez, J.A., do Nascimento, M.A., Macedo, J.B.M., Sliva, G.A., Oliveira, C.N., de Rezende, A.A., Brandao-Neto, J., Schwarz, A., and Almeida, M.G. 2012. Antioxidant activity and protective effect of *Turnera ulmifolia Linn. var. elegans* against carbon tetrachloride-induced oxidative damage in rats. *Food Chem. Toxicol.* 50(12): 4340–4347.

96. Banu, S., Bhaskar, B., and Balasekar, P. 2012. Hepatoprotective and antioxidant activity of *Leucas aspera* against D-galactosamine induced liver damage in rats. *Pharm. Biol.* 50(12): 1592–1595.

97. Huang, M., Jin, J., Sun, H., and Liu, G.T. 2008. Reversal of P-glycoprotein-mediated multidrug resistance of cancer cells by five schizandrins isolated from the Chinese herb Fructus Schizandrae. *Cancer Chemother. Pharmacol.* 62(6): 1015–1026.

98. Lin, Z., Huang, C.F., Liu, X.S., and Jiang, J. 2011. in vitro anti-tumor activities of quinolizidine alkaloids derived from *Sophora flavescens* Ait. *Basic Clin. Pharmacol. Toxicol.* 108(5): 304–309.

99. Wang, L., Li, F., Lu, J., Li, G., Li, D., Zhong, X.B., Guo, G.L., and Ma, X. 2010. The Chinese herbal medicine *Sophora flavescens* activates pregnane X receptor. *Drug Metab. Dispos.* 38(12): 2226–2231.

100. Zhang, Y., Liu, H., Jin, J., Zhu, X., Lu, L., and Jiang, H. 2010. The role of endogenous reactive oxygen species in oxymatrine-induced caspase-3-dependent apoptosis in human melanoma A375 cells. *Anticancer Drugs.* 21(5): 494–501.

101. Ho, J.W., Ngan Hon, P.L., and Chim, W.O. 2009. Effects of oxymatrine from Ku Shen on cancer cells. *Anticancer Agents Med. Chem.* 9(8): 823–826.

102. Zheng, H., Chen, G., Shi, L., Lou, Z., Chen, F., and Hu, J. 2009. Determination of oxymatrine and its metabolite matrine in rat blood and dermal microdialysates by high throughput liquid chromatography/tandem mass spectrometry. *J. Pharm. Biomed. Anal.* 49(2): 427–433.

4 Plant-Derived Active Phytochemicals Show Various Biologic and Pharmacological Activities in Cancer

Indole-3-carbinol and 3,3'-diindolylmethane are present in plants and medicinal herbs. These compounds exerted anticancer effects on prostate cancer cells through the inhibition of androgen- and estrogen-mediated pathways [1], and also showed inductive effects on xenobiotic metabolism. Both compounds are cyclin inhibitors and induced cell cycle arrest through the inhibition of insulin-like growth factor receptor 1 expression. However, both compounds showed different modes of action and efficacies due probably to a differential effect on binding affinity with mechanisms both dependent and independent of androgen receptor and aryl hydrocarbon receptor. Broccoli-derived phytochemicals also show pleiotrophic effects on cell growth and signaling pathways associated with cell proliferation. The results suggest that indole-3-carbinol and 3,3'-diindolylmethane can be developed as chemoprotective agents for cancer therapy.

MODULATOR OF AKT SIGNALING PATHWAY

Some of the phytochemicals from cruciferous vegetables are known to exert remarkable anticancer effects and produce health benefits. Sulforaphane, a phytochemical present in cruciferous vegetables showed inhibitory effects on phase II enzyme expression and it modulated the Akt signaling pathway in rat cardiomyocytes [2].

HERBAL EXTRACT–DRUG INTERACTIONS IN COMBINATION TREATMENT

An herbal extract is more active than the major phytochemicals it contains. Individual phytochemicals isolated from the active fraction may not show medicinal properties. The respective phytochemicals often form complexes

TABLE 4.1
Effects of Saponins on Cytochrome p450 Enzymes

1. CYP1A2
2. CYP2A6
3. CYP2B6
4. CYP2C9
5. CYP2E1
6. CYP3A4

in the active fractions, which can interact remarkably with cancer drugs to produce synergistic effects. However, the mode of actions of herbal extract–drug interactions is complex. An example with Rhizoma Paridis extract showed anticancer activity against lung cancer and liver cancer [3]. A combination of Rhizoma Paridis Saponins with cyclophosphamide was shown to attenuate the toxicity of cyclophosphamide. The study indicated the inhibitory effects of Rhizoma Paridis Saponins on the activity of cytochrome p450 enzymes as shown in Table 4.1 [3].

Rhizoma Paridis Saponins significantly inhibited the protein expression of CYP2B6 and CYP3A4 but showed little effects on CYP1A2, CYP2A6, CYP2C9, and CYP2E1 in rats. The results suggest that Rhizoma Paridis Saponins attenuated the transformation of cyclophosphamide to form active metabolites due to the inhibition of CYP2B6 and CYP3A4, which are believed to play an important role in the mediation of herb–drug interactions. The study provides insights into the enhancement of anticancer activity of cancer drugs that can be used in combination therapy with Rhizoma Paridis and other herbal medicines.

ENHANCEMENT OF CYTOTOXICITY OF *BETA VULGARIS* L.

Although some herbal medicines can enhance cytotoxicity of cancer drugs [4,5], clinical application is hampered by the undesirable side effects, probably due to herb–drug interactions, produced by the combination strategy. However, some of the herb–drug interactions could produce health benefits. Red beetroot (*B. vulgaris* L.) extract was approved by the Food and Drug Administration and European Union for medical use. The red food color E162, an active constituent of red beetroot extract was shown to reduce multiorgan tumor formations in an animal study [5]. The study implicated that red beetroot extract was safe to use as a chemopreventive agent in food supplements for humans. The combination of red beetroot extract and doxorubicin showed an enhanced cytotoxicity of the cancer drug in human cancer cell lines, suggesting that the combination of red beetroot extract and doxorubicin produced a synergistic activity. The in vivo study in the rat indicated that the cancer therapy with doxorubicin in combination with an appropriate herbal medicine could be optimized [5]. The synergistic cytotoxicity was maximal when the composition of red beetroot extract and doxorubicin was 1:5 in PaCa cells and MCF-7 cell lines. The findings infer that the composition of red beetroot extract and doxorubicin is a promising combination strategy for the treatment of pancreatic and breast cancers. However, further investigation is warranted to ensure the safety issue with long-term administration of herbs in combination with anticancer drugs.

HERB–DRUG INTERFERENCE IN CANCER TREATMENT

Malignant breast neoplasms are among the most frequent forms of cancer in the Western world. The common treatment method for breast cancer includes surgery, hormonal therapy, chemotherapy, radiation, and/or immunotherapy, yet these conventional treatments are often accompanied by severe side effects. Complementary and alternative medicine (CAM) treatments have been shown to be effective in alleviating those symptoms and/or increasing patient survival rates. Many CAM methods have been established to treat the patient in a holistic manner with consideration of the patient's psychological and spiritual needs.

The use of certain CAM methods may become problematic when patients experience undesirable side effects. Herbal medicines and dietary supplements, especially, may interfere with primary cancer treatments due to herb–drug interactions. Experienced physicians and Chinese medicine practitioners should coordinate in the treatment methods to optimize the effectiveness of the integrated approach; thus, integrative oncology makes a holistic approach to cancer care possible. This treatment strategy is beneficial to patients. The possible implications for the practice of integrative medicine can provide useful insights into integrative medicine for cancer therapy.

Numerous studies suggest integrative medicines are a promising alternative for the treatment of cancer. An early study reported that patients with breast cancer who sought integrative medical treatment when prescribed with tamoxifen or trastuzumab for target therapy showed noticeable health benefits. The Chinese medicinal formula, Si-Wu-Tang (SWT) was shown to stimulate MCF-7 cell growth by the activation of estrogen receptor α and human epidermal growth factor receptor 2 signaling [6]. SWT demonstrated herb–drug interference with cell proliferation in tumor-bearing mice in vivo and the attenuation of proliferation capacity in breast cancer cells in vitro. The integrative treatment with SWT and tamoxifen reversed tamoxifen-induced antiproliferative activity with an increase in estrogen receptor α and N-cadherin expression in MCF-7 cell line. Moreover, SWT reversed trastuzumab-induced antiproliferative activity in human epidermal growth factor receptor 2 cell lines, including SK-BR-3 and BT-474 through increased phosphorylation of the cell cycle regulatory proteins p27 (kip1) and p38, the antiapoptosis protein. The study suggests more attention should be paid to the possible interference between some of the herbal medicines and anticancer drugs [7].

CYP3A4-MEDIATED HERB–DRUG INTERACTIONS

The use of herbal medicines for the treatment of cancer has become popular among cancer patients in Asia. Concomitant use of herbal medicines and therapeutic drugs could produce health benefits to cancer patients. However, the safety issues and the efficacy of the combination therapy with some herbal medicines remain sketchy. Herbal medicines have the potential to cause pharmacokinetic interactions with therapeutic drugs resulting in either decreased or increased plasma levels of therapeutic drugs. This could cause toxicities and reduction in efficacy of drugs in cancer patients [8,9]. Significant herb–drug interactions were shown between St. John's Wort and imatinib and irinotecan. Most herb–drug interactions involve a drug-metabolizing enzyme system with cytochrome P450 enzyme system. The inhibitory effect of herbal medicines on CYP3A4 activity and related gene expression was reported [9]. However, details of the herb–drug interactions and the relevance of the data from in vitro studies in vitro to clinical significance in cancer patients is lacking. Midazolam, which is the common model substrate for cytochrome CYP3A4, is used to evaluate herb–drug interactions between therapeutic drugs and various common herbal medicines. Other dietary food chemicals including milk thistle, garlic, and *Panax ginseng* in combination with therapeutic drugs were studied. An in vitro study with St. John's Wort and anticancer drugs showed that it inhibited CYP3A4 in clinical trials with midazolam, irinotecan, and imatinib. The results indicated the combination treatment decreased plasma levels of these anticancer drugs due probably to CYP3A4 induction. There was no effect on CYP3A4 after treatment with garlic extract in combination with midazolam and docetaxel in vitro and in vivo, whereas milk thistle and *P. ginseng* showed CYP3A4 inhibition in vitro. However, the clinical trials showed that the combination treatment with herbs did not cause significant pharmacokinetic interactions with midazolam, irinotecan, docetaxel, and imatinib. It was believed that reduced bioavailability of these therapeutic drugs could contribute to the poor interactions between herbs and the therapeutic drugs in clinical trials. However, the in vitro findings provide useful information on the potential drug interactions between herbal medicines such as *Hypoxis hemerocallidea*, *Viscum album*, and St. John's Wort and therapeutic drugs. The discrepancies between in vitro and clinical trials with *P. ginseng* and milk thistle suggest that different herbal medicines may produce undesirable effects on combination treatment with therapeutic drugs. Midazolam is the common model substrate for CYP3A4 and is a good example used to demonstrate herb–drug interactions in clinical trials. However, more clinical studies are needed to confirm the health benefits of combination therapy with herbs and therapeutic drugs for the treatment of cancers.

TRADITIONAL CHINESE MEDICINE ON CYP3A4 ACTIVITY

Traditional Chinese medicines have been used as adjuvant therapy to attenuate the undesirable, harmful effects of therapeutic anticancer drugs in China [10]. However, pharmacokinetic interactions between traditional Chinese medicines and anticancer drugs can produce significant consequences for cancer patients. The drug-metabolizing enzyme cytochrome P4503A4 is believed to contribute to the significant pharmacokinetic interactions between drugs and *Oldenlandia diffusa*, *Codonopsis tangshen*, *Rehmannia glutinosa*, and *Astragalus propinquus* in human colon adenocarcinoma-derived LS 180 cells [10]. The herbal extracts of these traditional Chinese medicines inhibited CYP3A4 in humans. *O. diffusa* and *R. glutinosa* significantly induced PXR-mediated CYP3A4. Concurrent use of *O. diffusa* and *R. glutinosa* could lead to the reduction of the efficacy of therapeutic drugs. The increased toxicity caused by CYP3A4 inhibition by *A. propinquus* and *C. tangshen* suggests herb–drug interactions could undermine the concomitant treatment with some of the herbal medicines and therapeutic drugs. More research on the clinical relevance of the herb–drug interactions caused by herbal medicines is required.

ETOPOSIDE INTERACTION WITH *ECHINACEA*

Echinacea is known to have immune-stimulating properties and is used as an herbal supplement for the treatment of infections and common cold or flu [7]. It was reported that *Echinacea* inhibited cytochrome P450 3A4 and topoisomerase II in vitro. Topoisomerase II is the molecular target used in the treatment of lung cancer. Etoposide was reported to be metabolized by CYP3A4, suggesting possible drug–herbal interactions between *Echinacea* and etoposide. A previous study [7] reported a lung cancer patient who undertook chemoradiation with ciplatin and etoposide in combination with *Echinacea* showed a significant increase of platelet count. After discontinued administration of *Echinacea*, the platelet count of the patient remarkably decreased to a nadir of 44×10^3/L. The results infer that *Echinacea* played an important role in the patient's thrombocytopenia. Etoposide and other chemotherapeutic agents that are known to be CYP3A4 substrates can attenuate the efficacy of cancer drugs when administered in combination with some of the herbal medicines.

INTERACTION OF *V. ALBUM* L. WITH CHEMOTHERAPEUTIC DRUGS

Herbal medicines provide an excellent source of complementary and alternative medicine for the treatment of cancer. When it is used in combination with chemotherapeutic drugs, possible interactions between herbs and drugs may affect the effectiveness of the combination therapy. *V. album* L. was shown to produce cytostatic and cytotoxic activity in combination with doxorubicin in different cancer cell lines [8]. Table 4.2 shows the cancer cell lines used in the combination treatment [8].

The findings showed that *V. album* L. extract did not exert inhibitory effects on chemotherapy-induced cytostasis and cytotoxicity in the different cancer cells lines [8]. The treatment of cancer cell lines with therapeutic drugs and *V. album* L. extract indicated that alteration of drug efficacy can happen due to interactions between therapeutic drugs and *V. album* L. extract. The study suggests that the resulting efficacy of a therapeutic drug depends on the type of herbal medicines used in combination with therapeutic drugs. Herb–drug interactions can produce health benefits when *V. album* extract L. is used in combination with therapeutic drugs for the treatment of different cancers in humans.

TABLE 4.2
Combination Treatment of Cancer Cell Lines with Therapeutic Drugs

	Cancer Cell Lines	Drugs Treatment with *Viscum album* Extract
1.	Human breast carcinoma HCC1937 HCC1143	Doxorubicin
2.	Pancreas adenocarcinoma PA-TU-8902	Gemcitabine
3.	Prostate carcinoma DU145	Docetaxel and Mitoxantrone
4.	Lung carcinoma NCI-H460	Docetaxel and cisplatin

REFERENCES

1. Wang, T.T., Schoene, N.W., Milner, J.A., and Kim, Y.S. 2014. Broccoli-derived phytochemicals indole-3-carbinol and 3,3′-diindolylmethane exerts concentration-dependent pleiotropic effects on prostate cancer cells: Comparison with other cancer preventive phytochemicals. *Mol. Carcinog.* 51: 244–256.

2. Leoncini, E., Malaguti, M., Angeloni, C., Motori, E., Fabbri, D., and Hrelia, S. 2011. Cruciferous vegetable phytochemical sulforaphane affects phase II enzyme expression and activity in rat cardiomyocytes through modulation of Akt signaling pathway. *J. Food Sci.* 76: H175–H181.

3. Man, S., Li, Y., Fan, W., Gao, W., Liu, Z., Zhang, Y., and Liu, C. 2014. Combination therapy of cyclophosphamide and Rhizoma Paridis Saponins on anti-hepatocarcinoma mice and effects on cytochrome p450 enzyme expression. *Steroids* 80: 1–6.

4. Kapadia, G.J., Rao, G.S., Ramachandran, C., Iida, A., Suzuki, N., and Tokuda, H. 2013. Synergistic cytotoxicity of red beetroot (*Beta vulgaris* L.) extract with doxorubicin in human pancreatic, breast and prostate cancer cell lines. *J. Complement. Integr. Med.* 10(1): 113–122. doi: 10.1515/jcim-2013-0007.

5. Kan, S., Cheung, W.M., Zhou, Y., and Ho, W.S. 2014. Enhancement of doxorubicin cytotoxicity by tanshinone IIA in HepG2 human hepatoma cells. *Planta Med.* 80: 70–76.

6. Chen, J.L., Wang, J.Y., Tsai, Y.F., Lin, Y.H., Tseng, L.M., Chang, W.C., King, K.L., Chen, W.S., Chiu, J.H., and Shyr, Y.M. 2013. In vivo and in vitro demonstration of herb-drug interference in human breast cancer cells treated with tamoxifen and trastuzumab. *Menopause* 20: 646–654.

7. Bossaer, J.B. and Odle, B.L. 2012. Probable etoposide interaction with *Echinacea*. *J. Diet Suppl.* 9: 90–95.

8. Weissenstein, U., Kunz, M., Urech, K., and Baumgartner, S. 2014. Interaction of standardized mistletoe (*Viscum album*) extracts with chemotherapeutic drugs regarding cytostatic and cytotoxic effects in vitro. *BMC Complement. Altern. Med.* 14: 6.

9. Goey, A.K.L., Mooiman, K.D., Beijnen, J.H., Schellens, J.H., and Meijerman, I. 2013. Relevance of in vitro and clinical data for predicting CYP3A4-mediated herb-drug interactions in cancer patients. *Cancer Treat. Rev.* 36: 773–783.

10. Lau, C., Mooiman, K.D., Maas-Bakker, R.F., Beijnen, J.H., Schellens, J.H., and Meijerman, I. 2013. Effect of Chinese herbs on CYP3A4 activity and expression in vitro. *J. Ethnopharmacol.* 149: 543–549.

5 Therapeutic Benefits of Phytochemicals

Medicinal plants of different origins have been used for the treatment of diseases. Medicinal plants used by Himalayan for the treatment of hypertensions and uterine diseases are used as an anticoagulant [1]. Many of the medicinal properties of *Paeonia emodi* have been demonstrated in different animal models. Different classes of active phytochemicals such as triterpenoids, phenolics, and tannins are believed to be responsible for pharmacological activities [1]. These phytochemicals exhibit wide therapeutic activities, including free radical scavenging properties. Many of the natural products have been used to intervene in multistage carcinogenesis [2]. Different analogues based on plant-derived constituents have been prepared to evaluate the medicinal properties of potent phytochemicals. Studies have shown that natural product research has surged in the last two decades. Phytochemicals such as berberine, curcumin from turmeric, and genistein from soy bean have produced overwhelming publications. Among the natural antioxidants, ascorbic acid has been long believed to contribute to cancer therapy. The use of ascorbic acid in cancer treatment provides an alternative approach to cancer therapy. The mechanism of ascorbic acid for cancer therapy is attributed to its antioxidant properties and its ability to alter the metabolism of carcinogens, to enhance collagen synthesis essential for tumor encapsulation, and to stimulate the immune system. It is believed that mobilization of the immune system by anticancer agents can prevent cancer development due to interference with cancer cell signaling [2] and can fight against cancer too.

Amazonian people utilize a variety of active phytochemicals as a remedy for various diseases, including cognitive deficits in schizophrenia and dementias [3]. Approximately 300 Amazonian species with folk uses were shown in a database constructed from literature search and various psychoactive drug screening programs. Potential psychoactivity of phytochemicals was correlated to folk use of Amazonian psychoactive plant products. The results suggest the novelty and the potential of therapeutic active compounds for central nervous system. Many alternative and complementary medicines have been used for the treatment and management of dementia. Folk medicine such as *Medicago* is the common genus of the family Leguminosae with medicinal properties. Studies revealed that the major classes of phytochemicals of *Medicago sativa* (Linn) are saponins, flavonoids, terpenes, and coumarins, which exhibit multiarrays of pharmacological activities, including hypolipemic, antioxidant, and neuroprotective effects. *M. sativa* (Linn) appears to have excellent potential for developing new formulations in folk medicine with enhanced therapeutic benefits [4].

TREATMENT AND PREVENTION OF DEMENTIA WITH PHYTOCHEMICALS

Various chronic disorders of mental processes such as Alzheimer's disease have reached epidemic proportions, yet they cannot be satisfactorily managed by symptomatic treatment. There are only five drugs available for the management of cognitive symptoms and psychological symptoms of dementia [5]. Two of these drugs, both cholinesterase inhibitors, are derived from herbal phytochemicals to yield galantamine and rivastigmine. These active phytochemicals and their derivatives are used to treat dementia. The potential and therapeutic strategies with phytochemicals together with active herbal extracts for dementia have stimulated extensive research to develop new cholinesterase inhibitors. Alkaloid physostigmine from the calabar bean (*Physostigma venenosum*) has been used as a role model for the development of analogues that inhibit acetylcholinesterase. Other alkaloids including huperzine A from *Huperzia serrata* and galantamine from *Galanthus woronowii* were reported to be able to improve cognitive activities in Alzheimer's patients. Other phytochemicals such as cannabinoids from *Cannabis* saliva have become potential therapeutic agents for behavioral and psychological symptoms of dementia. In addition, resveratrol, which is present in various plants, is believed to exhibit pharmacological activities relevant to dementia. Curcumin from *Curcuma longa* also showed potential effects on delaying dementia progression. However, studies require more evidence for their clinical efficacy and drug safety. Other common medicinal plants include *Ginkgo biloba*, which has shown positive effects on cognitive functions in dementia patients, yet more consistent clinical studies are needed to confirm its therapeutic effectiveness (see Table 5.1).

In addition, Oleo gum resin that is secreted by *Commiphora mukul* has been used as an anticancer drug [6]. Oleo gum resin consisted of a mixture of essential oils, terpenes, polyphenols, flavanones, sterones, and Z- and E-guggulsterones. The active components have been identified in Oleo gum resin. Its mode of actions is through binding to nuclear receptors, leading to the modulation of the protein expression involved in carcinogenesis. Guggulsterones exhibited regulatory control over transcription factors such as nuclear factor (NF)-kappa B, and activator of transcription (signal transducer and activator of transcription [STAT]) and steroid

TABLE 5.1

Plant Extracts Show Promising Efficacy in Dementia Patients

Plant	Cognitive Functions
Saffron (*Crocus sativus*)	Cognitive symptoms
Curcumin (*Curcuma longa*)	Cognitive symptoms
Ginseng (*Panax* species)	Behavioral and psychological symptoms
Sage (*Salvia* species)	Cognitive symptoms
Lemon balm (*Melissa officinalis*)	Cognitive symptoms
Huperzine A (*H. serrata*)	Cholinesterase inhibitors
Galantamine (*Galanthus woronowii*)	Cholinesterase inhibitors
Cannabinoids (*Cannabis saliva*)	Cognitive symptoms

receptors. In addition, Oleo gum resin was reported to show different pharmacological activities. The study provides insights into the potential of natural products and their metabolites that can be useful as alternative strategies for prevention and treatment of cancer.

Many active phytochemicals with medicinal properties are found in vegetables, fruits, and plants. The phytochemicals obtained from terrestrial plants such as isoflavones are proven to produce health benefits. Some of these bioactive isoflavones can reduce the risk of cardiovascular diseases [7]. The cardioprotective effects of various phytochemicals are due to their antioxidative and anti-inflammatory activities and inhibition of platelet aggregation. The studies have provided leads for cardiovascular drug design. The use of natural products with different pharmacological activities for the treatment of various diseases and for drug development has become a common practice.

Fabaceae is the species commonly used in food supplements for the treatment of diseases in East Asia [8]. The studies showed that tannins, flavonoids, and some polyphenols in the plant extracts contributed to the pharmacological activities, including antifungal activities in cancer cell lines [8,9].

Metabolism of natural products is believed to be responsible for the pharmacological effects; thus, the studies on metabolism are of importance. It helps researchers to better understand the transformation process. The pharmacological actions and metabolic pathways need to be characterized to get a better understanding of the pharmacokinetics of natural products. Evidence-based assay methods are helpful in evaluating the effectiveness of natural products from active fractions.

Most natural product research is focused on the health benefits in vivo studies and is mostly market-driven studies. *Melicoccus bijugatus*, a member of the Sapindaceae, was reported to exhibit therapeutic benefits toward gastrointestinal disorders [9]. The pharmacological activities of

M. bijugatus fruits were formed to be different from other Sapindaceae species. The presence of specific phenolic compounds or sugar residues in *M. bijugatus* fruits is believed to contribute to the differences in medicinal properties of Sapindaceae fruits. The studies provide insights into predicting medicinal properties and food interactions that may be useful for evaluating the health benefits of other bijugatus fruits. The increasing incidence of drug resistance in cancer therapy has drawn considerable attention of the pharmaceutical and scientific communities toward studies in the potential anticancer activity of plant-derived substances, which are an excellent source of anticancer chemotypes. Herbal medicines are known to provide a rich source for drug development in pharmaceutics. The potent natural products derived from traditional Chinese medicines (TCMs) are believed to be associated with a wide variety of secondary metabolites. These metabolites are useful as alternative medicines for treating different diseases, including cancer. Active natural products are considered an untapped source of new drugs, and these natural products would undergo ingenious screening programs and would become useful therapeutic tools.

For many decades, the use of synthetic chemicals for cancer therapy has been effective. Many herbal medicines play an important role in health care and are used as alternative medicines for cancer therapy. The active phytochemicals from herbal medicines act as a defense system to fight against diseases. The multiarrays of medicinal properties of active phytochemicals are mediated by its structure–function relationship, which is important in new drug design. The use of TCMs for the treatment of ailments, including cancer, has become a common practice in Asian countries. TCMs and their natural products contain hundreds of chemically different phytochemicals, but only a few components in the herbal extract are potent. Berberine is a quaternary ammonium salt from the protoberberine group of isoquinoline alkaloids. It is found in plants such as *Berberis* (e.g., *Coptis chinensis* [Chinese goldthread or Huang Lian Su]) and is used as an antimicrobial and anti-inflammatory agent [10]. Clinical study showed berberine exhibited various pharmacological properties with superior therapeutic activities. Numerous studies have revealed that mixtures of phytochemical constituents produce therapeutic activities in the treatment of various ailments, cancers, diabetes, and hypercholesterolemia.

PHYTOCHEMICALS FOR CARDIOVASCULAR SYSTEM

The significance of natural products as therapeutic agents, especially those derived from herbal medicines, has been recognized in recent years. Intensive studies on the naturally occurring antispasmodic natural products for the treatment of cardiovascular diseases have been reported [11]. Hypertension poses a high risk for kidney and cardiovascular system. Recent study has revealed that treatment of hypertension can be significantly different among countries. A great deal of attention has focused on the high potential of antihypertensive agents as potential drugs. The research on naturally occurring antihypertensive agents is rapidly expanding. Among the antihypertensive phytochemicals, the high potential of diterpenes such as forskolin and stevioside exert vasorelaxant activity and vascular contractility [11]. Antihypertensive diterpenes can be used in the treatment of glaucoma that is characterized by the elevation of intraocular pressure.

TERPENOIDS AS PPAR MODULATORS

Medicinal plants contain many bioactive terpenoids, also known as isoprenoids. Terpenoids constitute one of the major families of phytochemicals with over 40,000 compounds. Some terpenoids such as artemisinin and taxol have found pharmaceutical applications as antimalarial and anticancer agents [11,12]. The bioactive terpenoids can modulate the activities of transcriptional factors such as peroxisome proliferator-activated receptors (PPARs). Terpenoids can be useful for the management of obesity-induced metabolic disorders, insulin resistance, and cardiovascular diseases.

ANTICANCER EFFECTS OF GAMBOGIC ACID

Some of the active principles from herbal medicines have been isolated and characterized for cancer drug development. Herbal medicine raises hope with its complementary role in the cancer treatment with cancer drugs. Gambogic acid (GA) is one of the naturally occurring compounds present in a brownish-to-orange resin called gamboge, which is derived from *Garcinia hanburyi* (Figure 5.1).

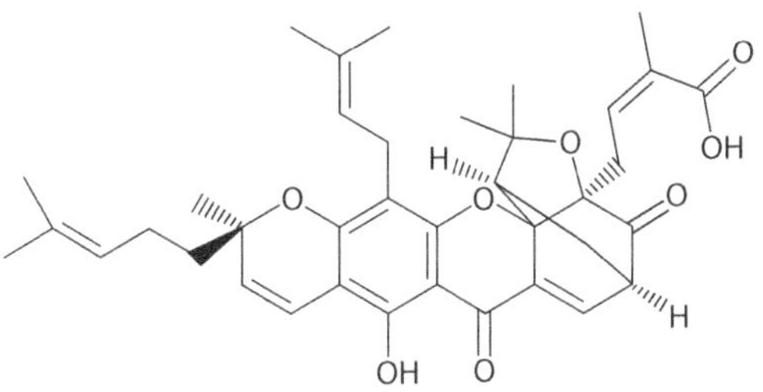

FIGURE 5.1 Chemical structure of gambogic acid.

G. hanburyi has a long history of medicinal use in Southeast Asia, and it is used as a folk medicine and coloring agent [13,14]. The anticancer activities of GA on two hepatocellular carcinoma cells with either p53 deletion (Hep3B) or p53 mutation (Huh7) were demonstrated in the previous study [13].

An improved separation method for the determination of 12 xanthones in gamboges from *G. hanburyi* enabled researchers to purify GA from *G. hanburyi* [15]. GA was reported to have potent antitumor activity via the induction of reactive oxygen species accumulation that consequently led to apoptosis of SMMC-7721 cells [16]. The mechanism of action of GA was believed to covalently modify I kappa B kinase-beta subunit to inhibit lipopolysaccharide-induced activation of NF-kappa B in macrophages [17]. GA mediated the control of nucleophosmin and nucleoporins in programmed cell death of Jurkat cells [18]. GA caused microtubule depolymerization and phosphorylation of c-Jun N-terminal kinase-1, leading to cell cycle arrest in MCF-7 cells [19]. The induction of apoptosis in cancer cells is vital in cancer treatment.

GA inhibited the growth of Hep3B and Huh7 through similar apoptotic pathway. After the treatment of Hep3B and Huh7 with GA for 24 h, IC_{50} was determined for both cell lines at 1.8 and 2.0 µM, respectively. The results showed that both cancer cells underwent morphological changes and DNA fragmentation. GA induced apoptosis in the two cell lines through caspase-3/7, caspase-8, and caspase-9 in the mitochondrial pathway. The results suggest that both the caspases in extrinsic death receptor pathway and the mitochondria-dependent pathway are involved in the GA-induced cell apoptosis. The inhibitory effects of GA on Hep3B and Huh7 are independent of p53-associated pathway.

GA showed inhibition of two different types of liver cancer cells [13]. GA showed potency to develop specific anticancer agents against different types of liver cancer. GA induced apoptosis via mitochondrial pathway in two types of liver cancer cells. p53 protein is a crucial factor in cellular stress responses and acts as an essential tumor suppressor [20]. Upon activation, p53 controls the signaling process associated with the cell cycle based on the severity of the DNA damage. Thus, it can inhibit cell cycle progression or induces apoptosis. Over 50% of human tumors have been reported to have p53 mutations, which affect p53 function. GA induced caspase-associated apoptotic pathway, which is important for the induction of caspase-regulated apoptosis. The two hepatocellular carcinoma (HCC) cell lines differ in the expression of one tumor suppressor protein, p53. Hep3B is with deleted p53, while Huh7 has mutated p53. Both cells are p53 deficient. p53, also known as tumor protein 53, is a tumor suppressor protein that in humans is encoded by the *TP53* gene. p53 plays a crucial role in the regulation of the cell cycle. The functions of the tumor suppressor gene have become an important target to prevent cancer. As such, p53 has been described as "the guardian of the genome" because of its pivotal role in conserving the stability of genome and preventing mutation [20]. p53 tumor suppressor is one of our defenses against damage due to radiation, carcinogens, and viruses. When DNA damage occurs, p53 levels rise and initiate protective measures. p53 binds to many regulatory sites in the genome that consequently halts cell division through the process of programmed cell death, or apoptosis. The mediation of p53 provides a common drug target for cancer development. However, the regulation of cell cycle in p53-deficient cells may be through different apoptotic pathways. Recent study showed p53 can mediate apoptosis by the repression of human papillomavirus (HPV) oncogenes and upregulation of tumor suppressor proteins in human cancer cells [21]. Hep3B contains an integrated hepatitis B virus (HBV) genome in the cell, which is highly associated with HCC development. GA can induce apoptosis in Hep3B and Huh7 cell death through caspases and is independent of p53-associated pathway. The results suggest the operative role of GA may involve killing of HBV genome integrated in HCC through regulation of other apoptotic processes [22]. These findings infer that the apoptosis of liver cancer is regulated by different complex systems involving large numbers of molecules and that GA-mediated signaling is only part of the regulatory system. The existence of the p53-independent regulatory signals in apoptosis in liver cancer cells may represent alternate approach for the inhibition of cancer cell growth to avoid complete liver damage. It was found that both the death receptor pathway and the mitochondrial pathway were involved in GA-induced apoptosis [13]. Several studies showed that Bid is an important pro-apoptotic protein in cross-linking the extrinsic cell death receptor signaling pathway to mitochondria upon caspase- 8-mediated cleavage [23]. The results are consistent with the previous study in that the integration of Bax and tBid to the outer membrane followed by the release of cytochrome c is a possible prerequisite for mitochondrial apoptotic pathway [24,25]. GA induces apoptosis in p53-deficient cancer cells. GA exerts inhibitory effects on Hep3B and Huh7 cell lines through similar mode of action. The study showed the anticancer activity of GA is mediated via both the caspases in extrinsic death receptor pathway and the mitochondria-dependent pathway. The present study has revealed that deletion or deficient of p53 does not affect cell cycle. The antitumor activity of GA may represent one of the molecular mechanisms involved in anticancer agent–induced apoptosis. The low concentration of GA toward cancer cells is one of the fundamental criteria for efficient drug development and targeting. GA may be an excellent antitumor agent for drug development.

The flavonoid silymarin is a phytochemical found in the seeds of "milk thistle" (*Silybum marianum*), which has been widely used as folk medicine in China. Silymarin is obtained from *Silybum marianum* (milk thistle), an edible plant that has been used for the treatment of liver-related disorders [26,27]. It exerts anticancer activity. Its mechanism of action includes inhibition of hepatotoxin binding to receptor sites on the hepatocyte membrane; reduction of glutathione oxidation to enhance its level in the liver and intestine; antioxidant activity; and stimulation of ribosomal RNA polymerase and subsequent protein synthesis, leading to enhanced hepatocyte regeneration. It is orally absorbed but has very poor bioavailability due to its poor water solubility.

The effects of silymarin on trichloroethylene (TCE) metabolism were demonstrated in the study [27]. Silymarin changed the level of metabolite formation and the selected hepatic enzyme activity in the liver, including CYP 2E1 and CYP 1A1. A direct correlation between TCE activation by P450s and its toxicity was established. CYP 2E1 and CYP 1A1 were both found to be responsible for the conversion of TCE to hydroxylated TCE and its corresponding acid. The severity of TCE-induced hepatotoxicity in the rat can be mediated by the treatment of rats with silymarin [26,27]. The experimental evidence showed that untreated rats displayed physiological changes in the liver, manifested as increased plasma concentrations of the liver enzymes, GOT and GPT.

Early studies have demonstrated that P450-mediated TCE metabolism produced hydroxylated TCE and its corresponding acid, which are believed to be responsible for the subsequent TCE-induced carcinogenicity. Selective hydroxylation of TCE is more closely associated with TCE-induced hepatotoxicity than GST-mediated metabolic pathway. Treatment of the rat with silymarin enhanced the detoxifying enzyme activities. The increase in GST activity resulted in the prevention of hydroxylated and acid form of TCE in covalent binding of macromolecules and the TCE-induced hepatotoxicity. It is possible that the activation of TCE could be mediated in part by GST, with significantly lower affinity for TCE than CYP 1A1 and CYP 2E1, which are the major enzymes associated with TCE metabolism.

TOXICITY OF SILYMARIN

Studies on the acute toxicity of silymarin after intravenous infusion have been carried out in mice, rats, rabbits, and dogs. The LD_{50} values were 400 mg/kg (mice), 385 mg/kg (rats), and 140 mg/kg (rabbits and dogs) [28]. These data demonstrate that the acute toxicity of silymarin is very low. Similarly, its subacute and chronic toxicities are also very low [29,30].

HEPATOPROTECTIVE ACTIVITIES

Carbon tetrachloride is known for its hepatotoxic properties and used for the evaluation of the pharmacological effects of silymarin. Silymarin has been shown to prevent carbon tetrachloride–induced lipid peroxidation and hepatotoxicity [31,32]. This effect of silymarin is attributed to its ability to normalize the levels of the transaminases that are elevated in hepatotoxicity [33].

ATTENUATION OF OXIDATIVE STRESS

Tert-butyl hydroperoxide has been found to induce microsomal lipid peroxidation and has been used as the model in different studies, demonstrating the protective effect of silymarin. Valenzuela and Guerra [34] demonstrated that silymarin inhibited oxygen consumption by rat microsomes, while Davila et al. showed that silymarin reduced enzyme loss and morphological alterations in neonatal rat hepatocytes [35]. Farghali et al. demonstrated the inhibition of lipid peroxidation by silymarin in perfused rat hepatocytes [36]. Other studies showed the protective effect of silymarin against ethanol-induced changes in these parameters [37,38].

MECHANISM OF ACTION OF SILYMARIN

Silymarin's hepatoprotective effects are purportedly accomplished via the following mechanism of actions:

- Antioxidation [39,40]
- Stimulation of ribosomal RNA polymerase and subsequent protein synthesis [41]
- Enhanced liver detoxification via the inhibition of phase I detoxification [42,43]
- Enhanced glucuronidation and replenishment of glutathione depletion [44]
- Inhibition of leukotriene and prostaglandin synthesis, Kupffer cell inhibition [30,45–47]
- Inhibition of cyclin-dependent kinases and arrest of cancer cell growth
- Immunomodulatory effects on the diseased liver [48,49]

Silymarin shows multiarrays of pharmacological activities and makes it a very promising drug of natural origin with little toxicity.

DOXORUBICIN

Doxorubicin is a widely used anticancer drug for the treatment of hemato-logic malignancies and solid cancer [50,51]. However, clinical use of doxo-rubicin is greatly limited by its serious adverse cardiac effects that result in cardiomyopathy and heart failure. Increased free radical production and reduced level of myocardial antioxidants are believed to be associated with the cardiotoxicity of doxorubicin [52,53]. Therefore, the development of strategy for therapies with doxorubicin is needed to reduce its cardiotox-icity. Chemoprotective agents are good candidates for clinical use with doxorubicin. They can be used to attenuate the cardiotoxicity of antitumor drugs.

CHLOROGENIC ACID

Chlorogenic acid (Figure 5.2) is a naturally occurring active phenolic com-pound found in many foods and botanical drugs. Due to its easy accessi-bility in human diet, numerous studies have been conducted to investigate its biological functions, including antioxidant, antihypertensive, anti-bacterial, anti-inflammatory, antifungal, antivirus, and anticarcinogenic activities [54–57].

DANSHEN

Danshen is also known as *Salvia miltiorrhiza*. Its dried root is commonly used in TCM for the treatment of heart disease, cerebrovascular disease, hepatitis, hepatocirrhosis, and cancer [58–60]. Some of the active phyto-chemicals can be isolated and purified from water-soluble and lipid-sol-uble fractions from *S. miltiorrhiza*. Salvianolic acid B (Figure 5.3) and tanshinone IIA (Figure 5.4) were considered as the quality elements for *S. miltiorrhiza*. Salvianolic acid B has multiarrays of pharmacological

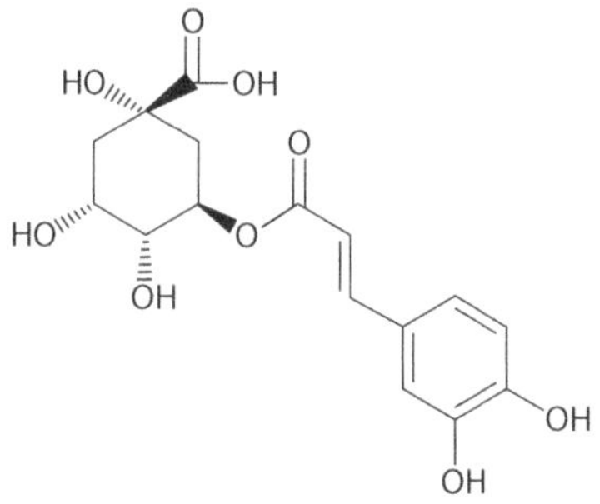

FIGURE 5.2   Chemical structure of chlorogenic acid.

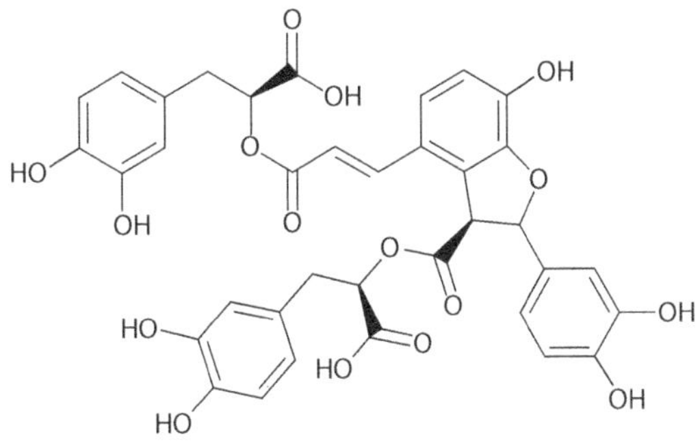

FIGURE 5.3   Chemical structure of salvianolic acid B.

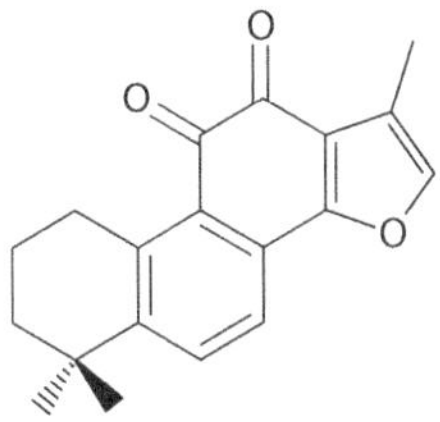

FIGURE 5.4   Chemical structure of tanshinone IIA.

effects on the liver and is a potential chemopreventive agent for head and neck squamous cell cancer [61–64]. Previous studies reported that tanshinone IIA exhibits cardiovascular activities including vasorelaxation, cardioprotective effects, and antitumor activity in many kinds of cancers including drug-resistant cancer cells [65–68].

The cytotoxic activity of doxorubicin was attenuated by chlorogenic acid and salvianolic acid B, which significantly improved the viability of doxorubicin-treated HepG2 cells [69]. The viability of HepG2 cells after treatment with a mixture of doxorubicin and chlorogenic acid and salvianolic acid B was increased to 89.2% and 91.2%, respectively, compared with the control. However, the cytotoxic activity of doxorubicin was enhanced by tanshinone IIA.

Lactate dehydrogenase (LDH) leakage, which was a marker of necrotic cellular death, has shown that HepG2 cells after treatment with doxorubicin resulted in the leakage of LDH. Tanshinone IIA increased the leakage of LDH. However, chlorogenic acid and salvianolic acid B cause the decrease of the cytotoxic effects of doxorubicin and resulted in the decrease of the LDH leakage compared to the doxorubicin-treated cancer cells.

The results showed that doxorubicin-treated cells markedly increased the activity of caspase-3 compared with the control and significantly enhanced the caspase-3 activity, leading to the death of HepG2 cells. Consequently, it killed the cancer cells with the LDH leakage.

The antioxidative activity of tanshinone IIA was measured using 2,2-diphenyl-1-picrylhydrazyl (DPPH), a stable free radical [70]. The scavenging activity of tanshinone IIA was lower than that of chlorogenic acid and salvianolic acid B. The results suggest the level of ROS would attribute the cytotoxic properties of doxorubicin in HepG2 cell line.

It is believed that the excess amount of ROS generation in tumor cells caused cardiac toxicity [52]. Polyphenols compounds could protect the heart through suppression of oxidative stress [71,72]. However, tanshinone IIA could attenuate the cardiac toxicity of doxorubicin through interactions with doxorubicin.

The elimination of ROS would reduce the cytotoxic properties of doxorubicin. Tanshinone IIA induced apoptosis and inhibited the proliferation and migration of cancer cells [73–75]. The present findings provide supportive evidences on its pharmacological properties.

Drug combination is most widely used in treating cancer. The Chou–Talalay method was most commonly used to assess the combination effect of two drugs [76,77]. The combination of doxorubicin and tanshinone IIA was assessed by combination index (CI). The CI value >1 was considered as synergism, CI value = 1 was considered as additive effect, and CI value >1 was considered as antagonism. The synergistic effects of the combination of doxorubicin and tanshinone IIA in HepG2 cell line were demonstrated [50].

The present findings suggest that the combination treatment may produce health benefits in cancer patients. *S. miltiorrhiza* is a popular TCM used for the treatment of cardiovascular disease and heart disease in China [78]. However, the administration of *S. miltiorrhiza* aqueous extract may attenuate the cytotoxic activity of doxorubicin. It is necessary to administer a suitable cardioprotective agent for chemotherapy with doxorubicin.

Activation of p53-independent pathway plays an important role in phytochemical-induced apoptosis and serves as a crucial factor for the invasion and metastasis of cancer. Liver metastasis and its treatment are related to the characteristics of the tumor and the immune systems of the host. Understanding the apoptotic pathways and their corresponding inhibitors enables us to formulate a suitable strategy for cancer therapy. The overexpression of genes that is associated with cell growth and programmed cell death depends on the activity of the p53-related pathways. During malignant progression, the human papillomaviruses integrate into the liver cell genome resulting in the loss of expression of oncogenes. The gene proteins may lead to interfere with the tumor suppression proteins p53. The tumor suppressor protein is activated in response to stress, and it plays an important role in regulation of cell cycle, DNA repair, and apoptosis [79]. Genomic alterations of p53 can be found in cancers. Deletion of p53 in cancer cells could lead to their resistance to apoptosis [79]. Tumor suppressor p53 is involved in transcriptional activation of the human *bax* gene [80,81]. Apoptosis-inducing factor played a role in the regulation of caspase-independent cell death [82,83]. The most common risk factor for the hepatocellular carcinoma is the HBV or hepatitis C viral (HCV) infection. There is a high incidence of HCV infection in Asia [84]. Antiviral approach against human transcriptional inactivation of viral infection is used for HCV infection. Complementary medicine using herbal ingredients against transcriptional inactivation of cancer cells showed minimal system toxicity and could be a promising agent for liver cancer therapy, particularly in an early stage of liver carcinogenesis. Cancer patients can have a higher survival rate with the complementary treatment.

EFFECTS OF *STEPHANIA TETRANDRA* IN COMBINATION WITH MEDICINAL HERBS EXHIBITED BENEFICIAL EFFECT IN CANCER PATIENTS

Tetrandrine (Figure 5.5) is a bisbenzylisoquinoline alkaloid, a naturally occurring compound isolated from the root of *Stephania tetrandra*, which was reported to exhibit a variety of pharmacological properties including anti-inflammatory, antirheumatic, and antihypertensive effects [85]. It can inhibit the proliferation of HeLa cells and HepG2 cells in vitro and suppress ascites tumors in mice [86]. Tetrandrine was reported to suppress Wnt/beta-catenin signaling and tumor growth of human colorectal cancer [87]. A previous study showed that tetrandrine induced apoptosis by activating reactive oxygen species and repressing Akt activity in human liver cancer cells [88]. The results suggest that mediation of ROS/Akt

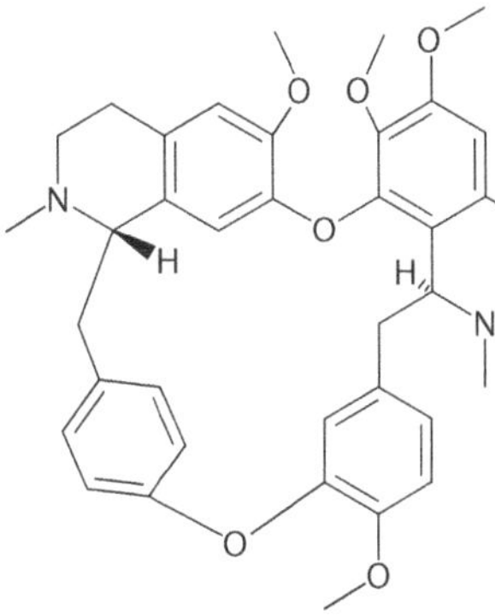

FIGURE 5.5   Chemical structure of tetrandrine.

pathway by tetrandrine can enhance the beneficial effects of tetrandrine in cells. Other study showed that tetrandrine was used together with cisplatin to enhance the growth suppression of ovarian cells and apoptosis [89]. In vivo study of the combined effects of tetrandrine and cisplatin exhibited anticancer effects in the rat [89]. The effect of tetrandrine with radiation on human esophageal cancer cell line TE1 showed that the expression of cyclin B1 protein increased while radiation-induced G2 arrest was abrogated [90]. The study suggests that the enhanced cytotoxicity and activation of ROS-dependent caspase-3 activity could induce programmed cell death in cancer [91,92]. Apoptosis of cancer cells was believed to be associated with the mitochondrial release of inducing factors that occurs downstream of cytochrome c release in response to oxidative stress [93,94].

Tetrandrine, an active component of *S. tetrandra*, was reported to have anticancer properties in cancer cells. However, the details of tetrandrine in liver cancer have not been elucidated yet. In this study, we investigated the effects of tetrandrine in hepatocarcinoma cells. The results showed that tetrandrine inhibited hepatocarcinoma cells' proliferation by suppression cell cycle progression at the G2/M phase. The change in the expression level of Bax, Bcl, p53, survivin, PCNA, poly (ADP-ribose) polymerase (PARP), and p21 was recorded. In addition, tetrandrine increased the caspase-3 expression and induced DNA fragmentation in Huh7 cells. The results suggest that the anticancer effect of tetrandrine in Huh7 may be mediated by p53-independent pathway.

The cell viability of Huh-7 decreased in a dose-dependent manner but remained basically unchanged in WRL68, the normal liver cells [95]. IC_{50} decreased from 20.8 to 8.0 μM for Huh-7 cells with time. The low value of IC_{50} suggests that tetrandrine can be a good candidate as an anticancer agent. The viability of WRL68 cells dropped and leveled off around 80.0% at 50 μM. No IC_{50} for WRL68 could be determined from the MTT assay. The results suggest tetrandrine is one of those few phytochemicals that can differentially act against cancer cell viability.

CELL DEATH OF HUH-7 CELLS INDUCED BY TETRANDRINE-MEDIATED THROUGH APOPTOSIS

DNA ladders were detected in the tetrandrine-treated Huh-7 cells [95]. The results provide supporting evidence that tetrandrine induces apoptosis of Huh-7 cells with increased concentration of tetrandrine. It implies that more genomic DNA molecules were cleaved to form smaller DNA fragments at a higher tetrandrine concentration. Huh-7 cell line is a p53 gene mutated cell line. In the absence of the functional p53 protein, apoptosis can still occur in Huh-7 cells, suggesting that apoptosis does not require the activation of p53 gene.

TETRANDRINE INDUCES G_1 PHASE CELL CYCLE ARREST

Studies show that there is a linkage between cell proliferation and apoptosis of Huh-7 [95]. The cell cycle analysis by flow cytometry shows that the percentage of cells in S phase and G_2 phase decreased. G_1 phase increased with increasing concentration of tetrandrine at all the time points. Tetrandrine induced programmed cell death via the mediation of p21 and proliferating cell nuclear antigen (PCNA) gene expression through probably the binding of p21 to cyclin–CDK 2 or cyclin–CDK 4 complexes and subsequently inhibited their activities. p21 plays an important role in cell cycle regulation by controlling the cell cycle progression from G_1 phase to S phase. PCNA is synthesized in the early G_1 phase and S phase during the cell cycle. It acts as an auxiliary factor for DNA polymerase δ in DNA synthesis during S phase of the cell cycle. It is an important protein responsible for the regulation of DNA synthesis. The binding of p21 to PCNA inhibits the role of PCNA during DNA replication. So the decreased protein expression for PCNA indicates that there were less cells entering S phase after treatment with tetrandrine. The results showed that a higher inhibitory effect on huh-7 cells was observed while proceeding from G_1 phase to S phase with higher concentration of tetrandrine. Therefore, there were less proliferating cells and larger population of cells was retained in G_1 phase after treatment. The data from reverse transcription polymerase chain reaction (RT-PCR) and Western blot also support the notion that tetrandrine induces G_1 phase cell cycle arrest.

TETRANDRINE-INDUCED APOPTOSIS INVOLVES THE INTRINSIC, CASPASE-DEPENDENT PATHWAY

Caspases are involved in both the initiation and execution of the programmed cell death. Western blot analysis showed that tetrandrine induced the apoptosis of Huh-7 through caspase activity. The expression for the cleaved, active caspase-9 was found to be increased. As caspase-9 is the initiator for the intrinsic apoptotic pathway, the cleavage of pro-caspase-9 to form the active caspase-9 is essential for inducing cell death in tetrandrine-treated Huh-7 cells and intrinsic apoptotic pathway is involved.

In the intrinsic apoptotic pathway, the caspase cascade involves active caspase-9 and procaspase-3. After the treatment of Huh-7 cells with tetrandrine, the protein expression for full-length PARP decreased in Huh-7 cells. The amount of cleaved PARP increased. PARP, an important protein for DNA repair, is the molecular substrate of active caspase-3. The occurrence of PARP cleavage is associated with DNA fragmentation in cells, resulting in cell death. In the gene expression analysis in Huh-7 cells, the survivin expression was found to be decreased in a concentration-dependent manner. It is reported that survivin is an antiapoptotic protein that exerts its function by binding to caspase-3 and hence the caspase-3 activity is suppressed. The decreased gene expression suggests that tetrandrine could promote apoptosis by suppressing the expression of survivin.

TETRANDRINE-INDUCED EXPRESSION OF PROTEINS IN BCL-2

The proapoptotic and antiapoptotic proteins participate in the apoptotic pathway associated with mitochondrial control. The gene expressions of Bax and Bid increased. Both the bax and bid are proapoptotic proteins where bcl-2 is an antiapoptotic protein. The pro- and antiapoptotic proteins exert their function in opposite ways. The elevated gene expressions of bax and bid, and the suppressed gene expression of bcl-2 suggest that these proteins were involved in apoptosis of Huh-7 cells.

REFERENCES

1. Zargar, B.A., Masoodi, M.H., Khanb, B.A., and Akbarc, S. 2013. *Paeonia emodi* Royle: Ethnomedicinal uses, phytochemistry and pharmacology. *Phytochem. Lett.* 6: 261–266.
2. Ullah, M.F., Bhat, S.H., Hussain, E., Abu-Duhier, F., Ahmad, A., and Hadi, S.M. 2012. Ascorbic acid in cancer chemoprevention: Translational perspectives and efficacy. *Curr. Drug Targets* 13: 1757–1771.
3. McKenna, D.J., Ruiz, J.M., Hoye, T.R., Roth, B.L., and Shoemaker, A.T. 2011. Receptor screening technologies in the evaluation of Amazonian ethnomedicines with potential applications to cognitive deficits. *J. Ethnopharmacol.* 134: 475–492.
4. Bora, K.S. and Sharma, A. 2011. Phytochemical and pharmacological potential of *Medicago sativa*: A review. *Pharm. Biol.* 49: 211–220.
5. Howes, M.J. and Perry, E. 2011. The role of phytochemicals in the treatment and prevention of dementia. *Drugs Aging* 28: 439–468.
6. Shah, R., Gulati, V., and Palombo, E.A. 2012. Pharmacological properties of guggulsterones, the major active components of gum guggul. *Phytother. Res.* 26: 1594–1605.
7. Vasanthi, H.R., ShriShriMal, N., and Das, D.K. 2012. Phytochemicals from plants to combat cardiovascular disease. *Curr. Med. Chem.* 19: 2242–2251.
8. Brito, S.A., Rodrigues, F.F., Campos, A.R., and da Costa, J.G. 2012. Evaluation of the antifungal activity and modulation between *Cajanus cajan* (L.) Millsp. leaves and roots ethanolic extracts and conventional antifungals. *Pharmacogn. Mag.* 8: 103–106.
9. Bystrom, L.M. 2012. The potential health effects of *Melicoccus bijugatus* Jacq. fruits: Phytochemical, chemotaxonomic and ethnobotanical investigations. *Fitoterapia* 83: 266–271.
10. Vuddanda, P.R., Chakraborty, S., and Singh, S. 2010. Berberine: A potential phytochemical with multispectrum therapeutic activities. *Expert Opin. Investig. Drugs* 19: 1297–1307.

11. Tirapelli, C.R., Ambrosio, S.R., de Oliveira, A.M., and Tostes, R.C. 2010. Hypotensive action of naturally occurring diterpenes: A therapeutic promise for the treatment of hypertension. *Fitoterapia* 81: 690–702.

12. Goto, T., Takahashi, N., Hirai, S., and Kawada, T. 2010. Various terpenoids derived from herbal and dietary plants function as PPAR modulators and regulate carbohydrate and lipid metabolism. *PPAR Res.* 2010: 483958.

13. Lee, P.N. and Ho, W.S. 2013. Antiproliferative activity of gambogic acid isolated from *Garcinia hanburyi* in Hep3B and Huh7 cancer cells. *Oncol. Rep.* 29(5): 1744–1750.

14. Asano, J., Chiba, K., Tada, M., and Yoshii, T. 1996. Cytotoxic xanthones from *Garcinia hanburyi*. *Phytochemistry* 3: 815–820.

15. Li, S., Song, J.Z., Han, Q.B., Qiao, C.F., and Xu, H.X. 2008. Improved high-performance liquid chromatographic method for simultaneous determination of 12 cytotoxic caged xanthones in gamboges, a potential anticancer resin from *Garcinia hanburyi*. *Biomed. Chromatogr.* 22: 637–644.

16. Nie, F., Zhang, X., Qi, Q., Yang, L., Yang, Y., Liu, W., Lu, N., Wu, Z., You, Q., and Guo, Q. 2009. Reactive oxygen species accumulation contributes to gambogic acid-induced apoptosis in human hepatoma SMMC-7721 cell. *Toxicology* 260: 60–67.

17. Palempalli, U.D., Gandhi, U., Kalantari, P., Vunta, H., Arner, R.J., Narayan, V., Ravindran, A., and Prabhu, K.S. 2009. Gambogic acid covalently modifies I kappa B kinase-beta subunit to mediate suppression of lipopolysaccharide-induced activation of NF-kappa B in macrophages. *Biochem. J.* 419: 401–409.

18. Shu, W., Chen, Y., Li, R., Wu, Q., Cui, G., Ke, W., and Chen, Z. 2008. Involvement of regulations of nucleophosmin and nucleoporins in gambogic acid-induced apoptosis in Jurkat cells. *Basic Clin. Pharmacol. Toxicol.* 103(6): 530–537.

19. Chen, J., Gu, H., Lu, N., Yang, Y., Liu, W., Qi, Q., Rong, J., Wang, X.T., You, Q.D., and Guo, Q.L. 2008. Microtubule depolymerization and phosphorylation of c-Jun N-terminal kinase-1 and p38 were involved in gambogic acid induced cell cycle arrest and apoptosis in human breast carcinoma MCF-7 cells. *Life Sci.* 83(3): 103–109.

20. Latonen, L. and Laiho, M. 2005. Cellular UV damage response-functions of tumor suppressor p53. *Biochim. Biophys. Acta* 1755: 71–89.

21. Munagala, R., Kausar, H., Munjal, C., and Gupta, R.C. 2011. Withaferin A induces p53-dependent apoptosis by repression of HPV oncogenes and upregulation of tumor suppressor proteins in human cervical cancer cells. *Carcinogenesis* 32(11): 1697–1705.

22. Yang, Y., Yang, L., You, Q.D., Nie, F.F., Gu, H.Y., Zhao, L., Wang, X.T., and Guo, Q.L. 2007. Differential apoptotic induction of gambogic acid, a novel anticancer natural product, on hepatoma cells and normal hepatocytes. *Cancer Lett.* 2: 259–266.

23. Li, H., Zhu, H., Xu, C., and Yuan, J. 1998. Cleavage of BID by caspase mediates the mitochondrial damage in the Fas pathway of apoptosis. *Cell* 94: 491–501.

24. Gross, A., Jockel, J., Wei, M.C., and Korsmeyer, S.J. 1998. Enforced dimerization of BAX results in its translocation, mitochondrial dysfunction and apoptosis. *EMBO J.* 14: 3878–3885.
25. Li, L., Lu, N., Dai, Q., Wei, L., Zhao, Q., Li, Z., He, Q., Dai, Y., and Guo, Q. 2011. GL-V9, a newly synthetic flavonoid derivative, induces mitochondrial-mediated apoptosis and G2/M cell cycle arrest in human hepatocellular carcinoma HepG2 cells. *Eur. J. Pharmacol.* 670(1): 13–21.
26. Dixit, N., Baboota, S., Kohli, K., Ahmad, S., and Ali, J. 2007. Silymarin: A review of pharmacological aspects and bioavailability enhancement approaches. *Indian J. Pharmacol.* 39(4): 172–179.
27. Song, J.Z. and Ho, W.S.J. 2014. Abri Cantongnesis modulates detoxifying enzymes to ameliorate hepatotoxicity in rats. *Altern. Integr. Med.* 3: 157.
28. Lecomte, J. 1975. Pharmacologic properties of silybin and silymarin. *Rev. Med. Liege.* 30: 110–114.
29. Morazzoni, P. and Bombardelli, E. 1994. *Silybum marianum (Carduus marianus)*. *Fitoterapia* 66: 3–42.
30. Saller, R., Meier, R., and Brignoli, R. 2001. The use of silymarin in the treatment of liver diseases. *Drugs* 61: 2035–2063.
31. Lettéron, P., Labbe, G., Degott, C., Berson, A., Fromenty, B., Delaforge, M., Larrey, D., and Pessayre, D. 1990. Mechanism for the protective effects of silymarin against carbon tetrachloride-induced lipid peroxidation and hepatotoxicity in mice. Evidence that silymarin acts both as an inhibitor of metabolic activation and as a chain-breaking antioxidant. *Biochem. Pharmacol.* 39: 2027–2034.
32. Muriel, P. and Mourelle, M. 1990. Prevention by silymarin of membrane alterations in acute carbon tetrachloride liver damage. *J. Appl. Toxicol.* 10: 275–279.
33. Sharma, A., Chakraborti, K.K., and Handa, S.S. 1991. Antihepatotoxic activity of some herbal formulations as compared to silymarin. *Fitoterapia* 62: 229–235.
34. Valenzuela, A. and Guerra, R. 1986. Differential effect of silybin on the Fe 2+ -ADP and t-butyl hydroperoxide-induced microsomal lipid peroxidation. *Experientia* 42: 139–141.
35. Davila, J.C., Lenherr, A., and Acosta, D. 1989. Protective effect of flavonoids on the drug-induced hepatotoxicity in vitro. *Toxicology* 57: 267–286.
36. Farghali, H., Kamenikova, L., and Hynie, S. 2000. Silymarin effects of intracellular calcium and cytotoxicity: A study in perfused rat hepatocytes after oxidative stress injury. *Pharmacol. Res.* 41: 231–237.
37. Wang, M., Grange, L.L., and Tao, J. 1996. Hepatoprotective properties of *Silybum marianum* herbal formulation on ethanol induced liver damage. *Fitoterapia* 67: 167–171.
38. Pepping, J. 1999. Milk thistle: *Silybum marianum*. *Am. J. Health Syst. Pharm.* 56: 1195–1197.
39. Mirguez, M.P., Anundi, I., Sainz-Pardo, L.A., and Lindros, K.O. 1994. Hepatoprotective mechanism of silymarin: No evidence for involvement of cytochrome P4502E1. *Chem. Biol. Interact.* 91: 51–63.

40. Wagner, H. 1981. Plant constituents with antihepatotoxic activity, in Beal, J.L. and Reinhard, E. (eds.) *Natural Products as Medicinal Agents*, Hippokrates-Verlag, Stuttgart, Germany.
41. Sonnenbichler, J. and Zetl, I. 1986. Biochemical effects of the flavonolignane silibinin on RNA, protein and DNA synthesis in rat livers. *Progr. Clin. Biol. Res.* 213: 319–331.
42. Baer-Dubowska, W., Szaefer, H., and Drajka-Kuzniak, V. 1998. Inhibition of murine hepatic cytochrome P450 activities by natural and synthetic phenolic compounds. *Xenobiotica* 28: 735–743.
43. Halim, A.B., el-Ahmady, O., Hassab-Allah, S., Abdel-Galil, F., Hafez, Y., and Darwish, A. 1997. Biochemical effect of antioxidants on lipids and liver function in experimentally-induced liver damage. *Ann. Clin. Biochem.* 34: 656–663.
44. Campos, R., Garrido, A., Guerra, R., and Valenzuela, A. 1989. Silybin dihemisuccinate protects against glutathione depletion and lipid peroxidation induced by acetaminophen on rat liver. *Planta Med.* 55: 417–419.
45. Saraswat, B., Visen, P.K.S., Patnaik, G.K., and Dhawan, B.N. 1995. Effect of andrographolide against galactosamine-induced hepatotoxicity. *Fototerapia* 66: 415–420.
46. Bosisio, E., Benelli, C., and Pirola, O. 1992. Effect of the flavanolignans of *Silybum marianum* L. on lipid peroxidation in rat liver microsomes and freshly isolated hepatocytes. *Pharmacol. Res.* 25: 147–154.
47. Dehmlow, C., Erhard, J., and de Groot, H. 1996. Inhibition of Kupffer cell functions as an explanation for the hepatoprotective properties of silibinin. *Hepatology* 23: 749–754.
48. Deak, G., Muzes, G., Lang, I., Niederland, V., Nıkαm, K., Gonzalez-Cabello, R., Gergely, P., and Fehér, J. 1990. Immunomodulator effect of silymarin therapy in chronic alcoholic liver diseases. *Orv Hetil.* 131: 1291–1292.
49. Lang, I., Nekam, K., Gonzalez-Cabello, R., Mūzes, G., Gergely, P., and Fehır, J. 1990. Hepatoprotective and immunological effects of antioxidant drugs. *Tokai J. Exp. Clin. Med.* 15: 123–127.
50. Schlumberger, M., Parmentier, C., Delisle, M.J., Droz, J.P., Sarrazin, D., and Couette, J.E. 1991. Combination therapy for anaplastic giant cell thyroid carcinoma. *Cancer* 67: 564–566.
51. Duggan, S.T. and Keating, G.M. 2011. Pegylated liposomal doxorubicin: A review of its use in metastatic breast cancer, ovarian cancer, multiple myeloma and AIDS-related Kaposi's sarcoma. *Drugs* 71: 2531–2558.
52. Singal, P.K., Li, T., Kumar, D., Danelisen, I., and Iliskovic, N. 2000. Adriamycin-induced heart failure: Mechanism and modulation. *Mol. Cell Biochem.* 207: 77–86.
53. Barry, E., Alvarez, J.A., Scully, R.E., Miller, T.L., and Lipshultz, S.E. 2007. Anthracycline-induced cardiotoxicity: Course, pathophysiology, prevention and management. *Expert Opin. Pharmacother.* 8: 1039–1058.
54. Bahadir, O., Citoglu, G.S., and Coban, T. 2010. Antioxidant activity of some *Scorzonera* species and quantitative analysis of chlorogenic acid. *Planta Med.* 76: 1227–1227.

55. Zhao, Y., Wang, J., Ballevre, O., Luo, H., and Zhang, W. 2012. Antihypertensive effects and mechanisms of chlorogenic acids. *Hypertens. Res.* 35: 370–374.

56. Wu, W.B., Hung, D.K., Chang, F.W., Ong, E.T., and Chen, B.H. 2012. Anti-inflammatory and anti-angiogenic effects of flavonoids isolated from *Lycium barbarum* Linnaeus on human umbilical vein endothelial cells. *Food Funct.* 3: 1068–1081.

57. Ozcelik, B., Kartal, M., and Orhan, I. 2011. Cytotoxicity, antiviral and anti-microbial activities of alkaloids, flavonoids, and phenolic acids. *Pharm. Biol.* 49: 396–402.

58. Lu, Y. and Foo, L.Y. 2002. Polyphenolics of *Salvia*—A review. *Phytochemistry* 59: 117–140.

59. Che, X.H., Park, E.J., Zhao, Y.Z., Kim, W.H., and Sohn, D.H. 2010. Tanshinone II A induces apoptosis and S phase cell cycle arrest in activated rat hepatic stellate cells. *Basic Clin. Pharmacol. Toxicol.* 106: 30–37.

60. Zhang, Y., Wei, R.X., Zhu, X.B., Cai, L., Jin, W., and Hu, H. 2012. Tanshinone IIA induces apoptosis and inhibits the proliferation, migration, and invasion of the osteosarcoma MG-63 cell line in vitro. *Anti-Cancer Drugs* 23: 212–219.

61. Lay, I.S., Chiu, J.H., Shiao, M.S., Lui, W.Y., and Wu, C.W. 2003. Crude extract of *Salvia miltiorrhiza* and salvianolic acid B enhance in vitro angiogenesis in murine SVR endothelial cell line. *Planta Med.* 69: 26–32.

62. Zhao, Y., Hao, Y., Ji, H., Fang, Y., Guo, Y., Sha, W., Zhou, Y., Pang, X., Southerland, W.M., Califano, J.A., and Gu, X. 2010. Combination effects of salvianolic acid B with low-dose celecoxib on inhibition of head and neck squamous cell carcinoma growth in vitro and in vivo. *Cancer Prev. Res.* 3: 787–796.

63. Wang, R., Yu, X.Y., Guo, Z.Y., Wang, Y.J., Wu, Y., and Yuan, Y.F. 2012. Inhibitory effects of salvianolic acid B on CCl(4)-induced hepatic fibrosis through regulating NF-kappaB/IkappaBalpha signaling. *J. Ethnopharmacol.* 144: 592–598.

64. Joe, Y., Zheng, M., Kim, H.J., Kim, S., Uddin, M.J., Park, C., Ryu, do G., Kang, S.S., Ryoo, S., Ryter, S.W., Chang, K.C., and Chung, H.T. 2012. Salvianolic acid B exerts vasoprotective effects through the modulation of heme oxygenase-1 and arginase activities. *J. Pharmacol. Exp. Ther.* 341: 850–858.

65. Su, C.C., Chen, G.W., Kang, J.C., and Chan, M.H. 2008. Growth inhibition and apoptosis induction by tanshinone IIA in human colon adenocarcinoma cells. *Planta Med.* 74: 1357–1362.

66. Hong, H.J., Liu, J.C., Chen, P.Y., Chen, J.J., Chan, P., and Cheng, T.H. 2012. Tanshinone IIA prevents doxorubicin-induced cardiomyocyte apoptosis through Akt-dependent pathway. *Int. J. Cardiol.* 157: 174–179.

67. Jiang, B., Zhang, L., Wang, Y., Li, M., Wu, W., Guan, S., Liu, X., Yang, M., Wang, J., and Guo, D.A. 2009. Tanshinone IIA sodium sulfonate protects against cardiotoxicity induced by doxorubicin in vitro and in vivo. *Food Chem. Toxicol.* 47: 1538–1544.

68. Slusarczyk, S., Zimmermann, S., Kaiser, M., Matkowski, A., Hamburger, M., and Adams, M. 2011. Antiplasmodial and antitrypanosomal activity of tanshinone-type diterpenoids from *Salvia miltiorrhiza*. *Planta Med.* 77: 1594–1596.

69. Kan, S., Cheung, W.M., Zhou, Y., and Ho, W.S. 2014. Enhancement of doxorubicin cytotoxicity by tanshinone IIA in HepG2 human hepatoma cells. *Planta Med.* 80(1): 70–76.

70. Rao, Y.K., Geethangili, M., Fang, S.H., and Tzeng, Y.M. 2007. Antioxidant and cytotoxic activities of naturally occurring phenolic and related compounds: A comparative study. *Food Chem. Toxicol.* 45: 1770–1776.

71. Chlopcíková, S., Psotová, J., Miketová, P., Sousek, J., Lichnovský, V., and Simánek, V. 2004. Chemoprotective effect of plant phenolics against anthracycline-induced toxicity on rat cardiomyocytes. Part II. Caffeic, chlorogenic and rosmarinic acids. *Phytother. Res.* 18: 408–413.

72. Jiang, B., Zhang, L., Li, M., Wu, W., Yang, M., Wang, J., and Guo, D.A. 2008. Salvianolic acids prevent acute doxorubicin cardiotoxicity in mice through suppression of oxidative stress. *Food Chem. Toxicol.* 46: 1510–1515.

73. Liu, F., Yu, G., Wang, G., Liu, H., Wu, X., Wang, Q., Liu, M., Liao, K., Wu, M., Cheng, X., and Hao, H. 2012. An NQO1-initiated and p53-independent apoptotic pathway determines the anti-tumor effect of tanshinone IIA against non-small cell lung cancer. *PLoS One.* 7: e42138.

74. Tang, C., Xue, H.L., Bai, C.L., and Fu, R. 2011. Regulation of adhesion molecules expression in TNF-alpha-stimulated brain microvascular endothelial cells by tanshinone IIA: Involvement of NF-kappaB and ROS generation. *Phytother. Res.* 25: 376–380.

75. Lee, W.Y., Cheung, C.C., Liu, K.W., Fung, K.P., Wong, J., Lai, P.B., and Yeung, J.H. 2010. Cytotoxic effects of tanshinones from *Salvia miltiorrhiza* on doxorubicin-resistant human liver cancer cells. *J. Nat. Prod.* 73: 854–859.

76. Chou, T.C. and Talalay, P. 1984. Quantitative-analysis of dose-effect relationships: The combined effects of multiple-drugs or enzyme-inhibitors. *Adv. Enzyme Regul.* 22: 27–55.

77. Chou, T.C. 2010. Drug combination studies and their synergy quantification using the Chou-Talalay method. *Cancer Res.* 70: 440–446.

78. You, J.S., Pan, T.L., and Lee, Y.S. 2007. Protective effects of Danshen (*Salvia miltiorrhiza*) on adriamycin-induced cardiac and hepatic toxicity in rats. *Phytother. Res.* 21: 1146–1152.

79. Schuler, M., Bossy-Wetzel, E., Goldstein, J.C., Fitzgerald, P., and Green, D.R. 2000. p53 induces apoptosis by caspase activation through mitochondrial cytochrome c release. *J. Biol. Chem.* 275: 7337–7342.

80. Chipuk, J.E., Maurer, U., Green, D.R., and Schuler, M. 2003. Pharmacologic activation of p53 elicits Bax-dependent apoptosis in the absence of transcription. *Cancer Cell.* 4: 371–381.

81. Oda, E., Ohki, R., Murasawa, H., Nemoto, J., Shibue, T., Yamashita, T., Tokino, T., Taniguchi, T., and Tanaka, N. 2000. Noxa, a BH 3-only member of the Bcl-2 family and candidate mediator of p53-induced apoptosis. *Science* 288: 1053–1058.

82. Cregan, S.P., Fortin, A., MacLaurin, J.G., Callaghan, S.M., Cecconi, F., Yu, S.W., Dawson, T.M., Dawson, V.L., Park, D.S., and Kroemer, G. 2002. Apoptosis-inducing factor is involved in the regulation of caspase-independent cell death. *J. Cell Biol.* 158: 507–517.

83. Miyashita, T. and Reed, J.C. 1995. Tumor suppressor p53 is a direct transcriptional activator of the human *bax* gene. *Cell* 80: 293–299.

84. Jemal, A., Bray, F., Center, M.M., Ferlay, J., Ward, E., and Forman, D. 2011. Global cancer statistics. *CA: Cancer J. Clin.* 61(2): 69–90.

85. Ng, L.T., Chiang, L.C., Lin, Y.T., and Lin, C.C. 2006. Antiproliferative and apoptotic effects of tetrandrine on different human hepatoma cell lines. *Am. J. Chin. Med.* 34(1): 125–135.

86. Yoo, S.M., Oh, S.H., Lee, S.J., Lee, B.W., Ko, W.G., Moon, C.K., and Lee, B.H. 2002. Inhibition of proliferation and induction of apoptosis by tetrandrine in HepG2 cells. *J. Ethnopharmacol.* 81(2): 225–229.

87. He, B.C., Gao, J.L., Zhang, B.Q., Luo, Q., Shi, Q., Kim, S.H., Huang, E., Gao, Y., Yang, K., Wagner, E.R., Wang, L., Tang, N., Luo, J., Liu, X., Li, M., Bi, Y., Shen, J., Luther, G., Hu, N., Zhou, Q., Luu, H.H., Haydon, R.C., Zhao, Y., and He, T.C. 2011. Tetrandrine inhibits Wnt/beta-catenin signaling and suppresses tumor growth of human colorectal cancer. *Mol. Pharmacol.* 79(2): 211–219.

88. Liu, C., Gong, K., Mao, X., and Li, W. 2011. Tetrandrine induces apoptosis by activating reactive oxygen species and repressing Akt activity in human hepatocellular carcinoma. *Int. J. Cancer* 129(6): 1519–1531.

89. Zhang, Y., Wang, C., Wang, H., Wang, K., Du, Y., and Zhang, J. 2011. Combination of tetrandrine with cisplatin enhances cytotoxicity through growth suppression and apoptosis in ovarian cancer in vitro and in vivo. *Cancer Lett.* 304(1): 21–32.

90. Yu, J., Liu, F., Sun, M., Sun, Z., and Sun, S. 2011. Enhancement of radiosensitivity and the potential mechanism on human esophageal carcinoma cells by tetrandrine. *Cancer Biother. Radiopharmacol.* 26(4): 437–442.

91. Li, X., Zhen, D., Lu, X., Xu, H., Shao, Y., Xue, Q., Hu, Y., Liu, B., and Sun, W. 2010. Enhanced cytotoxicity and activation of ROS-dependent c-Jun NH2-terminal kinase and capase-3 by low doses of tetrandrine-loaded nanoparticles in lovo cells – A possible Trojan strategy against cancer. *Eur. J. Pharmaceut. Biopharmaceut.* 75(3): 334–340.

92. Arnoult, D., Parone, P., Mattinou, J.C., Antonsson, B., Estaquier, J., and Ameisen, J.C. 2002. Mitochondrial release of apoptosis-inducing factor occurs downstream of cytochrome c release in response to several proapoptotic stimuli. *J. Cell Biol.* 159: 923–929.

93. Susin, S.A., Daugas, E., Ravagnan, L., Samejima, K., Zamzami, N., Loeffler, M., Costantini, P., Ferri, K.F., Irinopoulou, T., Prévost, M.C., Brothers, G., Mak, T.W., Penninger, J., Earnshaw, W.C., and Kroemer, G. 2000. Two distinct pathways leading to nuclear apoptosis. *J. Exp. Med.* 192: 571–580.

94. Li, X., Su, B., Liu, R., Wu, D., and He, D. 2011. Tetrandrine induces apoptosis and triggers caspase cascade in human bladder cancer cells. *J. Surg. Res.* 166(1): E45–E51.

95. Yu, V.W. and Ho, W.S. 2013. Tetrandrine inhibits hepatocellular carcinoma cell growth through the caspase pathway and G2/M phase. *Oncol. Rep.* 29(6): 2205–2210.

6 Mechanism of Cancer Drug Action

Cell growth and mitochondria have become intriguing targets for anticancer agents, inherent to a vast majority of tumors. Herbal phytochemicals that target cell growth factors and mitochondria have emerged to become a focus of drug research due to their remarkable potential in clinical applications. The potential of cell cycle regulation as a target for anticancer agents has been reinforced by the encouraging findings that show promise in a number of cancer mutations (Figure 6.1). Cancers should be treated by agents that target multiple genes or pathways. Cancer is unlikely to be treated by single drugs that target only a single gene or pathway. This is in line with the notion that we should define a class of anticancer drugs acting on the control of cell growth and mitochondria. The targets on cell growth regulation are of major importance from the point of view of their role in uncontrollable cell growth by small natural products. Herbal phytochemicals, especially small molecules, hold a considerable promise as potential anticancer agents that can exert inhibitory effects on apoptosis. The common inhibition of apoptosis is shown in Figure 6.2.

INHIBITION OF APOPTOTIC ACTIVITY

Animal models that are able to reveal the action of herbal anticancer compounds in tumor can have important impact on drug development. Potential drugs can act on the vascularization without cytotoxic effects in vitro, yet many of the antiangiogenic phytochemicals show different pharmacological activities in animal studies. The antiangiogenic effects of phytochemicals were implemented based on inhibitory effects on the cancer growth. The mice model was tested with success on a number of experiments in tumor-bearing mice after the administration of the drug bevacizumab either alone or in combination with another investigational herbal phytochemicals. The model produced useful tumor growth data [1]. The information can be used for the novel drug design in preclinical experiments.

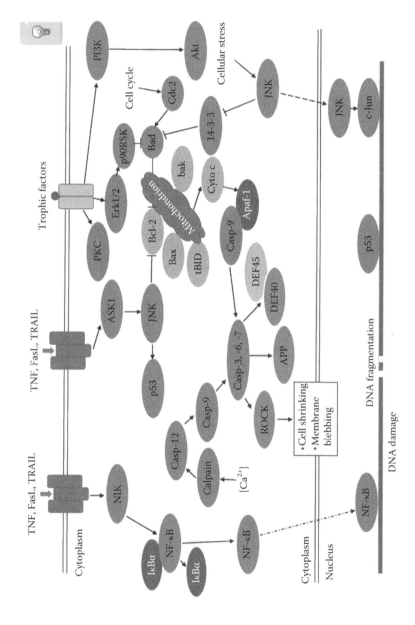

FIGURE 6.1 **(See color insert.)** Regulation of apoptosis.

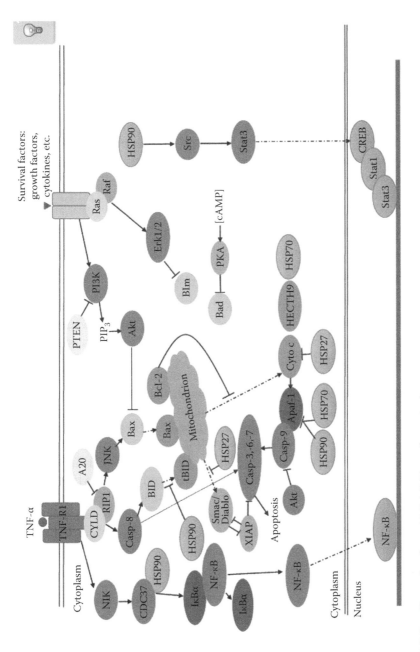

FIGURE 6.2   (See color insert.) Inhibition of apoptosis.

GPR30/EG FG SIGNALING PATHWAY

Cancer cell proliferation can be assayed by different labeling reagents, including 5′-bromo-2′-deoxyuridine in primary cancer cells. The expression of proliferation and apoptotic markers is analyzed by immunoblotting and the enzyme-linked immunosorbent assay (ELISA).

Nude mice bearing subcutaneous implanted-Ishikawa tumors were used in the study of the antitumorigenic action of antiestrogenic benzopyran derivative ($k - 1$). Compound $k - 1$ inhibited the proliferation of endometrial adenocarcinoma cells and decreased the expression of GPR30-regulated proteins, resulting in apoptosis by increasing the expression of apoptotic makers including NOXA and a novel proapoptotic gene (PUMA) alpha accompanying with a decrease in the expression of p-CREB and Bcl-xL. Herbal phytochemicals that interfered with GPR30-regulated epidermal growth factor receptor (EGFR) activation decreased p-ERK (extracellular signal-regulated kinases), p-c-jun, c-fos, cyclin D1, and c-myc expression. Compound $k - 1$ is a potent inhibitor on cancer cell growth [2].

RAF/MEK/ERK AND P13K/AKT PATHWAYS

The RAF/MEK/ERK and P13K/AKT pathways are highly implicated in the development of pancreatic cancer. The cytotoxic effects of cotreatment with Sorafenib, a RAF inhibitor, and HS-173, a P13K inhibitor, induced G2/M arrest and increased apoptosis with the loss of mitochondrial membrane potential [3]. Combined treatment of the two drugs synergistically inhibited the Panc-1 cell viability. The study demonstrated that apoptosis of cancer cells could result from the cotreatment of phytochemicals. The synergistic action was accompanied by increased levels of cleaved caspase-3 and poly (ADP-ribose) polymerase (PARP), which is the common mechanism of actions observed in human liver cancer cells after treatment with active phytochemicals [4,5]. Simultaneously targeting the RAF/MEK and P13K/AKT pathways can induce a synergistic inhibitory effect on pancreatic cancers. The cotreatment of anticancer phytochemicals may be considered to be a new therapeutic strategy for treating various cancers (Table 6.1).

Phytochemicals from herbal medicines, especially those with low formula weight, have become the mainstay of cancer chemotherapy for over two decades. Single chemical natural products of herbs origin and some of their synthetic derivatives are tested clinically as novel chemotherapeutic cancer drug candidates. Some of the bioactive phytochemicals from herbal medicines with scientific data can contribute to future anticancer drug discovery. However, some other phytochemicals may not have high potency in in vivo studies. Combination of active phytochemicals has become the mainstream of research.

TABLE 6.1
Akt Substrates Table

Substrate	Isoform	Substrate Function and Effect of Phosphorylation
Akt1	Akt1	Activated by insulin and various growth and survival factors to function in a wortmannin-sensitive PI3 kinase-involved pathway controlling survival and apoptosis. Autophosphorylation activates the kinase
B-Raf	Akt1	Signaling intermediate in Erk 1/2 pathway.
	Akt3	Phosphorylation causes inhibition.
BAD	Akt1	Proapoptotic protein. Phosphorylation inhibits function and promotes survival.
Bcl-xL	Akt1	Prevents apoptosis through binding to apoptotic proteins. Phosphorylation promotes VDAC binding.
Casp9	Akt1	Protease initiates apoptosis. Phosphorylation inhibits protease activity.
ER-α	Akt1	Nuclear receptor and transcription factor.
	Akt2	Phosphorylation activates the receptor and increases gene expression, causing mammary and uterine cell proliferation.
ER-β	Akt1	Nuclear receptor and transcription factor. Phosphorylation prevents cofactor binding and decreases activity.
GAPDH	Akt2	Catalyzes the phosphorylation of glyceraldehyde-3-phosphate during glycolysis. Phosphorylation decreases nuclear translocation and glyceraldehyde 3-phosphate dehydrogenase (GAPDH)-induced apoptosis.
GSK3α	Akt1	Serine/threonine protein kinase that phosphorylates and inactivates glycogen synthase. Phosphorylation inhibits activity.
GSK3β	Akt1	Serine/threonine protein kinase that phosphorylates and inactivates glycogen synthase. Phosphorylation inhibits activity.
HSP27	Akt1	Heat shock protein that confers cellular resistance to stress and adverse environmental change. Phosphorylation alters tertiary structure and modulates actin polymerization and reorganization.
MDM2	Akt1	Ubiquitin ligase involved in p53 degradation. Phosphorylation results in translocation to the nucleus and inhibition of p53.
MDM4	Akt1	RING-finger domain protein involved in p53 degradation and apoptosis. Phosphorylation stabilizes MDM4 and MDM2.
p21Cip1	Akt1	Regulates cell cycle and cell survival. Phosphorylation increases protein stability.

(Continued)

TABLE 6.1 (*Continued*)
Akt Substrates Table

Substrate	Isoform	Substrate Function and Effect of Phosphorylation
p27Kip1	Akt1	A cyclin-dependent kinase inhibitor that enforces the G1 cell cycle restriction point.
		Phosphorylation promotes 14-3-3 binding and cytoplasmic localization.
Raf1	Akt1	Signaling intermediate in Erk 1/2 pathway.
		Phosphorylation inhibits activity.
Ron	Akt1	RTK for macrophage stimulating protein, cell adhesion, proliferation, and migration.
		Phosphorylation causes 14-3-3 binding.
SOX2	Akt1	A transcription factor required for early embryogenesis and embryonic stem cell pluripotency.
		Phosphorylation stabilizes SOX2, increasing transcriptional activity.

STAT3 was validated as a novel anticancer drug target. STAT3 is a small molecule inhibitor that can play an important role in oncogenesis and metastasis. 2-Methoxystypandrone and anthraquinones are the active principles isolated from the roots of *Polygonum cuspidatum* by activity-guided assays. 2-Methoxystypandrone exhibited a significant inhibitory effect on STAT3 activation and cell proliferation of human breast cancer. The studies with quinone analogues suggest that the phenolic and carbonyl groups of the active principles contribute to the inhibitory activities in the STAT3 signaling [5,6].

INHIBITION OF TUMOR ANGIOGENESIS BY NATURAL PRODUCTS

Angiogenesis, which plays a pivotal role in the promotion of cancer, provides a promising molecular target to cancer therapy. Inhibitory phytochemicals for angiogenesis would deplete the cell oxygen and nutrition supply. Angiogenesis inhibitors can be used for cancer treatment in attenuating the vascular endothelial growth factor (VEGF) signal pathway. Small anticancer phytochemicals inhibited specifically receptor tyrosine kinases (RTKs). However, other affected RTKs may be involved in tumor growth [7]. The anticancer properties of these inhibitors can serve as templates for the development of novel small anticancer drugs. There is increasing evidence for the therapeutic effects of active phytochemicals on signaling pathways that are associated with cancer growth.

Polyphenolic compounds are the most common class of phytochemicals that show promise as angiogenesis inhibitors. Prominent herbal polyphenols include epigallocatechin gallate and genistein from soy bean. Their pharmacological effects on cancer warrant further studies.

ANTIPROLIFERATIVE ACTIVITY OF GAMBOGIC ACID

The anticancer activities of active phytochemicals may show inhibitory effects on cancer cell growth. Gambogic acid, which was isolated from *Garcinia hanburyi*, demonstrated antiproliferative activities in both Hep3B and Huh7 cancer cells [8]. The differences in antiproliferative activities in both cell lines suggest different modes of actions of gambogic acid. Hep3B have p53 deletion, while Huh7 have p53 mutation. After treatment with gambogic acid, both Hep3B and Huh7 cells underwent morphological changes and consequently resulted in apoptosis of cells. Gambogic acid induced apoptosis through caspase-3/7, caspase-8, and caspase-9 of the caspase cascade in the mitochondrial pathway. Both the caspases are involved in the extrinsic death receptor pathway (Figure 6.3) and the mitochondrial-dependent pathway (Figure 6.4) in apoptosis of cancer cells. However, the inhibitory activity of gambogic acid is independent of p53-associated pathway in Hep3B and Huh7 cells.

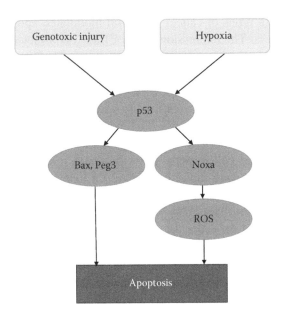

FIGURE 6.3 Modulation of extrinsic pathway by herbal medicines.

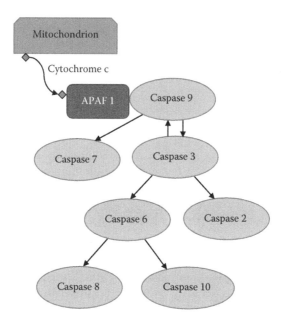

FIGURE 6.4   Herbal medicines can modulate intrinsic pathway (mitochondrial pathway).

The apoptosis of both types of cancer cells involved the activation of caspase-8 and subsequently led to Bid cleavage. Full-length Bid (24 kDa) was found to be cleaved and subsequent accumulation of tBid in the mitochondria in both types of cancer cells. Bid modulated apoptosis through signaling process. The expression level of Bax was affected in the apoptotic process. Gambogic acid induced apoptosis in Hep3B and Huh7 cells through caspase cascade independent of p53-mediated pathway. The biologic action of gambogic acid suggests that the apoptosis of liver cancer cells may be mediated by different complex systems involving other signaling pathways and receptors. The existence of the p53-independent regulatory signals in apoptosis may represent an alternate approach to inhibition of cancer cell growth. Both the death receptor and the mitochondrial pathways were found to be involved in gambogic acid–induced apoptosis. Bid plays an important role in crosstalk between the extrinsic cell death receptor signaling pathway and mitochondrial pathway via caspase-8-mediated cleavage [9]. The study showed that the anticancer activity of gambogic acid is mediated via both the caspases in the extrinsic death receptor pathway and the mitochondria-dependent pathway. The deletion or the mutant of p53 in cancer cells did not play a role in the apoptosis of Hep3B and Huh7 cells. Gambogic acid may be a promising therapeutic anticancer agent.

6-SHOGAOL INHIBITS CANCER CELL INVASION VIA THE BLOCKADE OF NUCLEAR FACTOR-κB ACTIVATION

Shogaols are the active constituents of gingers. They were shown to display anticancer activities in breast cancer cells. Among Shogaols, 6-, 8-, and 10-shogaol inhibited phorbol 12-myristate 13-acetate (PMA)-stimulated MDA-MB-231 cell invasion with a decrease in MMP-9 secretion [10]. 6-Shogaol displayed the most anti-invasive effects on MMP-9 gene activation with the subsequent reduction in NF-κB transcription activity via the inhibition of IκB phosphorylation and suppression of NF-κB p65 phosphorylation and nuclear translocation. 6-Shogaol was found to inhibit c-Jun N-terminal kinase (JNK) activation, ERK, and NF-κB signaling in PMA-stimulated MMP-9 activation. The molecular mechanism of action involves the downregulation of MMP-9 transcription via the NF-κB activation cascade. This type of small naturally occurring herbal phytochemicals show promise as antimetastatic antitumor agents.

ALBACONOL SUPPRESSES LPS-TRIGGERED CYTOKINE PRODUCTION

Herbs contain abundant natural products with different anticancer activities. Albaconol is a novel prenylated resorcinols isolated from the budding tissues of the *Albatrellus confluens*, inedible mushroom. Albaconol was shown to inhibit cancer cell growth through immunosuppression. Albaconol inhibited maturation and antigen presentation of dendritic cells [11]. Albaconol inhibited lipopolysaccharides (LPS)-induced production of proinflammatory cytokines TNF-alpha, IL-6, and IL-1 beta and expression of MHC-II and significantly inhibits T-cell-stimulating capacity of dendritic cells and antigen-specific T-cell response. Blockage of LPS-induced NF-kappa B activation led to immunosuppression or anti-inflammatory activity. Some of the small natural occurring phytochemicals with anticancer activity can be a potent immunosuppressive and anti-inflammatory agent. Their anticancer activity is through inhibition of dendritic cell function and NF-Kappa B activation.

METHYL ANGOLENSATE INDUCES MITOCHONDRIAL PATHWAY OF APOPTOSIS IN DAUDI CELLS

Small-size natural products from herbal medicines can be very potent in the treatment of various cancers. Methyl angolensate, a tetranor-triterpenoid that is present in the roots of callus of *Soymida febrifuga*, also known as Indian red wood tree, showed remarkable cytotoxicity in Burkitt's lymphoma cell lines [12]. After the treatment of Burkitt's lymphoma cells, methyl angolensate activated DNA double-strand break repair proteins including KU70 and KU80 in surviving cancer cells. The enhancement of cytotoxicity of methyl angolensate is through generation of reactive oxygen species and to enhance the loss of mitochondrial trans-membrane potential.

PROMISING ALKALOIDS FOR CANCER THERAPY

Other small-size natural products with anticancer properties include phenanthroindolizidine and phenanthroquinolizidine, which also showed antiamoebicidal and anti-inflammatory activity. No compounds in this class have passed clinical trials due to low bioavailability and limited cytotoxicity in central nervous system [13]. This class of compounds could suppress protein and nucleic acid synthesis. However, the molecular targets have not yet been identified. Nevertheless, an extensive effort has been made on the synthesis of the analogues in the hope that a novel anticancer agent can be developed. A novel anticancer agent with kinase affinity, MR22388, belongs to the tripentone family with high cytotoxicity, and this exerts in vitro cytotoxic activities against numerous cancer cell lines [14]. Tripentone compounds are commonly found in herbal medicines, yet its anticancer properties remain sketchy. The mode of action of MR22388 showed that it is a weak inhibitor of the polymerization of tubulin [15]. It includes apoptosis via the MAP kinase pathways. However, MR22388 can be a remarkable inhibitor of tyrosine kinase FLT3-ITD, which is a mutated form of the RTK FLT3. The inhibition of kinases resulted in the activation of the kinase in normal karyotypes' acute myeloid leukemia (AML), of which inhibitors have been extensively investigated. It has emerged as one of the major approaches to treatment stratification in normal karyotypes' AMLs. The potent cytotoxic tripentone showed its multiple targets involving both tubulin and kinases.

TARGETING THE NF-KAPPA B AND mTOR PATHWAYS

The presence of TNF-alpha in tumors suggests that NF-Kappa B and the mTOR pathways are activated. Inhibitor of I Kappa B kinase beta (IKK beta) can act as the signaling mode of checkpoint that regulates transcription via the p-I Kappa B alpha/NF-Kappa B axis and subsequently regulates translation via the mTOR/p-S6k/p-eIF4EBP axis. The analogue of quinoxaline 13-197 was identified as an IKK beta inhibitor in pancreatic cancer cells [16]. The quinoxaline 13-197 inhibited the kinase activity of IKK beta and TNF-alpha-mediated NF-Kappa B transcription in pancreatic cells with low potency (mmol/L). It also blocked the phosphorylation of I Kappa B alpha, S6k, and eIF4EBP, resulting in cell cycle arrest in the G1 phase. The quinoxaline 13-197 protected mice from lipopolysaccharide-induced cell death and inhibited metastasis in an orthotopic pancreatic cancer model.

QUINOLINE DERIVATIVES AGAINST BREAST CANCER

Camptothecin is a cytotoxic quinoline alkaloid isolated from the bark and stem of *Camptothecin acuminata,* which has been used for cancer treatment in China for over a century. Camptothecin classes of compounds are effective against a broad spectrum of tumors. Camptothecin inhibits human DNA topoisomerase I by blocking the cleavage/religation reaction of topoisomerase I, resulting in the accumulation of cleaved complex [17,18]. It is highly cytotoxic via collisions between replication forks and topoisomerase I cleavable complexes, resulting in the formation of covalent topo I–DNA complexes. The reactions involve activation of the ubiquitin/26S proteasome pathway and SUMO conjugation to topoisomerase I. The derivatives of camptothecin are potent anticancer agents with the similar mode of actions in cancer cells. An oxime, a derivative of camptothecin inhibited cell cycle progression and induced DNA polyploidy. Oxime induces caspase-3 activation, PARP cleavage, and gamma-H2A phosphorylation, resulting in apoptosis of breast cancer cells.

CHEMOPREVENTIVE POTENTIAL OF CURCUMIN

Curcumin, a diarylheptanoid, is the principal curcuminoid present in Asia spice turmeric, which belongs to Zingiberaceae (ginger family). The curcuminoids are natural polyphenols that can exist in several tautomeric forms. Curcumin modulates the inflammatory response by downregulating cycloxgenase-2, lipoxygenase, and nitric oxide synthase enzymes and inhibits several other enzymes involved in inflammation mechanisms [19,20]. A combination of aspirin, curcumin, and sulforaphane reduced cell viability through activation of caspase-3 and poly (ADP-ribose) polymerase proteins. The NF-Kappa B DNA binding activity was inhibited in MIA PaCa-2 and Panc-1 cells. The findings suggest that the combination of aspirin, curcumin, and sulforaphane inhibits cell growth by inducing cell apoptosis and sustained activation of the ERk1/2, c-Jun, and p53 signaling pathways.

PHARMACOLOGICAL ACTIVITY OF BERBERINE

Previous studies reported the pharmacological properties of *Coptis chinensis* Franch formulation for the treatment of various liver disorders

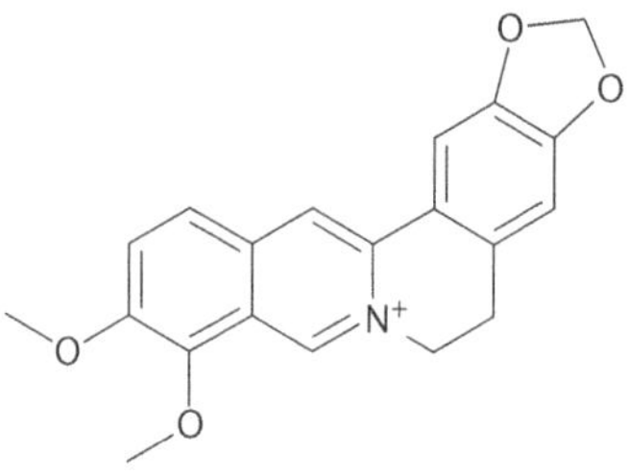

FIGURE 6.5 Chemical structure of berberine.

and its major constituent triterpenoids [21–23]. One of the major principles was demonstrated to exhibit marked anti-inflammatory and antiproliferative effects on cancer cells. The antitumor potential of berberine
(Figure 6.5), a naturally bioactive phytochemical from *Coptis chinensis*
Franch against Huh7 cancer cells and WRL 68 liver cells, was demonstrated [24]. The results showed berberine-induced apoptosis of liver
cancer cells through procaspase-9 and its effector caspases, procaspase-3
and procaspase-7. Flow cytometry showed that berberine exhibited cell
cycle arrest at the M/G1 phase. RT-PCR analysis showed that berberine
increased the expression of Bax, leading to activation of caspase cascade.
The present findings demonstrate that berberine induced apoptosis of Huh7
via mitochondrial pathway.

Berberine reduced cell viabilities in a dose-dependent manner [24]. It
exhibited relatively much less effect on normal liver cells. The results suggest that berberine exhibited differential effects on normal and cancer cells.
The difference in inhibitory activities suggests that berberine could initiate
different signaling processes in normal and cancer cells. FITC Annexin V
and PI costaining showed that berberine caused apoptosis in HCC cells.

The results confirmed that berberine induced apoptosis of cancer cells.
Bax, Bid, CIDEA, HRK, and p21 were found to be upregulated, while
AKT and Bcl-2 were found to be downregulated by berberine. The change
in the gene expression suggested that berberine induced apoptosis of Huh7
cell lines through their intrinsic proapoptotic activity that was reported to
be associated with mitochondrial dysfunction in cell death. AKT protein
was reported to participate in the regulation of tumor cell survival and
proliferation by stabilizing p21 protein [25]. Bcl-2 gene expression was
found to be decreased. The results suggest that berberine inhibited the
gene expression of Bcl-2 gene in HCC cells. The present results indicate
that survivin negatively regulated apoptosis by inhibiting caspase activation. The results reflected that an overexpression of Bcl-2 may cause
the accumulation of cells in the G0 phase of cell cycle division, resulting
in chemoresistance [26]. The effector caspases, caspase-7 and caspase-3,

are downstream targets of caspase-9. PARP is a well-known downstream target of active caspase-3 [27]. PARP was reported to produce poly (ADP-ribosyl)ation of nuclear proteins with NAD as substrate. PARP is inactivated by cleavage. The present findings showed that the protein expression of full-length PARP decreased while the cleaved form of which increased. Since berberine decreased the protein expression of procaspase-9 and its downstream effector caspases, procaspase-3 and procaspase-7, it is possible that berberine cleaves caspase-9, caspase-3, and caspase-7. Cleaved caspases become active executioners of the intrinsic apoptotic pathway. The protein expression of proliferating cell nuclear antigen (PCNA) was found to be downregulated by berberine in Huh7 cells. The experimental results imply that as the expression of PCNA was reduced by berberine, and fewer cells were able to repair damaged DNA. The results show that berberine possessed potent anticancer activities in human hepatocellular carcinoma.

REGULATION OF APOPTOSIS BY PHYTOCHEMICALS

Apoptosis is a regulated suicide mechanism characterized by nuclear condensation, cell shrinkage, membrane blebbing, and DNA fragmentation. Activity associated with caspases cascade involves a family of cysteine proteases [28–34]. Initiator caspases (including caspase-2, caspase-8, caspase-9, caspase-10, caspase-11, and caspase-12) are closely coupled to proapoptotic signals. Once activated, these caspases cleave and activate downstream effector caspases (including caspase-3, -6, and -7), which in turn execute apoptosis by cleaving cellular proteins following specific Asp residues. The activation of Fas and TNFR by FasL and TNF, respectively, leads to the activation of caspase-8 and -10. DNA damage induces the expression of primary immunodeficiency diseases (PIDD), which binds to death adaptor molecule (RAIDD) and caspase-2, and leads to the activation of caspase-2. Cytochrome c released from damaged mitochondria is coupled to the activation of caspase-9. X-linked inhibitor of apoptosis protein (XIAP) inhibits caspase-3, caspase-7, and caspase-9. Mitochondria release multiple proapoptotic molecules, such as Smac/Diablo, AIF, HtrA2, and EndoG, in addition to cytochrome c. Smac/Diablo binds to XIAP, preventing it from inhibiting caspases. Caspase-11 is induced and activated by pathological proinflammatory and proapoptotic stimuli and leads to the activation of caspase-1, thereby promoting inflammatory response and apoptosis by directly processing caspase-3. Caspase-12 and caspase-7 are activated under ER stress conditions. Antiapoptotic ligands, including growth factors and cytokines, activate Akt and p90RSK. Akt inhibits Bad by direct phosphorylation and prevents the expression of Bim.

INHIBITION OF APOPTOSIS BY PHYTOCHEMICALS

Cell survival requires the active inhibition of apoptosis via inhibiting the expression of proapoptotic factors as well as promoting the expression of antiapoptotic factors [29,31,35–38]. The activity of PI3K pathway can be modulated by anticancer phytochemicals, resulting in the activation of Akt, an important player in survival signaling. PTEN negatively regulates the PI3K/Akt pathway. Activated Akt phosphorylates, leading to inhibition of the proapoptotic Bcl-2 family members Bad, Bax, caspase-9, GSK-3, and FoxO1. Many growth factors and cytokines induce antiapoptotic Bcl-2 family members. The Jaks and Src phosphorylate and activate Stat3, which in turn induces the expression of Bcl-xL and Bcl-2. Erk1/2 and PKC activate p90RSK, which activates CREB and induces the expression of Bcl-xL and Bcl-2. These Bcl-2 family members protect the integrity of mitochondria, preventing cytochrome c release and the subsequent activation of caspase-9. TNF-α may activate both proapoptotic and antiapoptotic pathways; TNF-α not only can induce apoptosis by activating caspase-8 and caspase-10 but can also inhibit apoptosis via NF-κB, which induces the expression of antiapoptotic genes such as Bcl-2. cIAP1/2 inhibits TNF-α signaling by binding to TRAF2. FLIP inhibits the activation of caspase-8.

MITOCHONDRIAL CONTROL OF APOPTOSIS

The Bcl-2 family of proteins regulates apoptosis by controlling mitochondrial permeability. The antiapoptotic proteins Bcl-2 and Bcl-xL reside in the outer mitochondrial wall and inhibit cytochrome c release [39–42]. The proapoptotic Bcl-2 proteins Bad, Bid, Bax, and Bim may reside in the cytosol but translocate to mitochondria following death signaling, where they can be modulated by external stimuli or phytochemicals, leading to the release of cytochrome c (Figure 6.6). Bad translocates to mitochondria and forms a proapoptotic complex with Bcl-xL. This translocation is mediated by survival factors regulating the phosphorylation of Bad, leading to its cytosolic sequestration. Cytosolic Bid is cleaved by caspase-8 following signaling through Fas; its active fragment (tBid) translocates to mitochondria. Bax and Bim translocate to mitochondria in response to death stimuli, including survival factor withdrawal. Activated following DNA damage, p53 induces the transcription of Bax, Noxa, and PUMA. Upon release from mitochondria, cytochrome c binds to Apaf-1 and forms an activation complex with caspase-9. Although the mechanism(s) regulating mitochondrial permeability and the release of cytochrome c during apoptosis are not fully understood, Bcl-xL, Bcl-2, and Bax may influence the voltage-dependent anion channel (VDAC), which may play a role in regulating cytochrome c

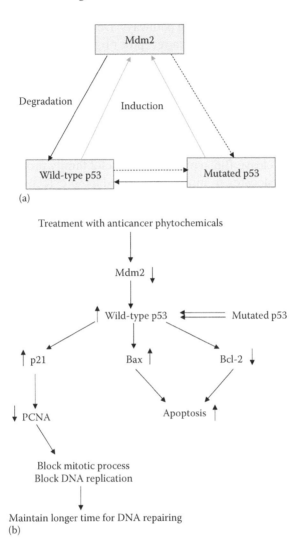

FIGURE 6.6 Modulation of apoptotic activities through p53-dependent apoptosis pathway. (a) Anticancer activity of phytochemicals on p53. (b) Effects of anticancer phytochemicals on p53-dependent pathway.

release. Mule/ARF-BP1 is a DNA damage–activated E3 ubiquitin ligase for p53 and Mcl-1, an antiapoptotic member of Bcl-2.

DEATH RECEPTOR SIGNALING PATHWAY

Apoptosis can be induced through the activation of death receptors including Fas, TNFαR, DR3, DR4, and DR5 by their respective ligands [29,43–45]. Some of the active phytochemicals are reported to effectively modulate the receptor signaling pathways. Death receptor ligands charac-teristically initiate signaling via receptor oligomerization, which in turn results in the recruitment of specialized adaptor proteins and activation of caspase cascades. Binding of FasL induces Fas trimerization, which recruits initiator caspase-8 via the adaptor protein Fas-associated protein with death domain (FADD). Caspase-8 then oligomerizes and is acti-vated via autocatalysis. Activated caspase-8 stimulates apoptosis via two parallel cascades: it can directly cleave and activate caspase-3, or alter-natively, it can cleave Bid, a proapoptotic Bcl-2 family protein. Truncated Bid (tBid) translocates to mitochondria, inducing cytochrome c release, which sequentially activates caspase-9 and -3. TNF-α and DR-3L can deliver pro- or antiapoptotic signals. TNFαR and DR3 promote apoptosis via the adaptor proteins TRADD/FADD and the activation of caspase-8. The interaction of TNF-α with TNFαR may activate the NF-κB pathway via NIK/IKK. The activation of NF-κB induces the expression of pro-survival genes including Bcl-2 and FLIP; the latter can directly inhibit the activation of caspase-8. FasL and TNF-α may also activate JNK via ASK1/MKK7. The activation of JNK may lead to the inhibition of Bcl-2 by phosphorylation. In the absence of caspase activation, the stimulation of death receptors can lead to the activation of an alternative programmed cell death pathway termed necroptosis by forming complex IIb.

CELL CYCLE G1/S CHECKPOINT SIGNALING PATHWAY

The primary G1/S cell cycle checkpoint controls the commitment of eukaryotic cells to transition through the G1 phase to S phase. Two cell cycle kinase complexes, CDK4/6-Cyclin D and CDK2-Cyclin E, work in concert to relieve the inhibition of a transcription complex that contains the retinoblastoma protein (Rb) and E2F [46–49]. In G1-phase uncommit-ted cells, hypophosphorylated Rb binds to the E2F-DP1 transcription fac-tors, forming an inhibitory complex with histone deacetylases (HDAC) to repress key downstream transcription events. Sequential phosphory-lation of Rb by Cyclin D-CDK4/6 and Cyclin E-CDK2 dissociates the HDAC-repressor complex, permitting the transcription of genes for DNA replication. Akt can phosphorylate FoxO1/3, which inhibits their func-tion by nuclear export, thereby allowing cell survival and proliferation. Importantly, different checkpoint controls, including TGF-β, DNA damage, replicative senescence, and growth factor withdrawal, can be modulated by phytochemicals and cancer drugs. Phytochemicals can act on transcription factors to induce specific members of the INK4 or Kip/Cip families of cyclin dependent kinase inhibitors (CKIs). Notably, the oncogenic poly-comb protein Bmi1 acts as a negative regulator of INK4A/B expression in stem cells and human cancer. At a critical convergence point with the

DNA damage checkpoint, cdc25A is ubiquitinated and targeted for degradation via the Skp, Cullin, F-box containing protein (SCF) ubiquitin ligase complex downstream of the ATM/ATR/Chk pathway. The growth factor withdrawal activates GSK-3β to phosphorylate cyclin D, which leads to its rapid ubiquitination and proteasomal degradation. Ubiquitin/proteasome-dependent degradation and nuclear export are mechanisms commonly used to effectively reduce the concentration of cell cycle control proteins and are explored as therapeutic targets for cancer treatment in humans.

REFERENCES

1. Rocchetti, M., Germani, M., Del Bene, F., Poggesi, I., Magni, P., Pesenti, E., and De Nicolao, G. 2013. Predictive pharmacokinetic modeling of tumor growth after administration of an anti-angiogenic agent, bevacizumab, as single-agent and combination therapy in tumor xenografts. *Cancer Chemother. Pharmacol.* 71: 1147–1157.

2. Chandra, V., Fatima, I., Saxena, R., Hussain, M.K., Hajela, K., Sankhwar, P., Roy, B.G., Chandna, S., and Dwivedi, A. 2013. Anti-tumorigenic action of 2-[piperidinoethoxyphenyl]-3-[4-hydroxyphenyl]-2H-benzo(b)pyran: Evidence for involvement of GPR30/EGFR signaling pathway. *Gynecol. Oncol.* 129: 433–442.

3. Yun, S.M., Jung, K.H., Lee, H., Son, M.K., Seo, J.H., Yan, H.H., Park, B.H., Hong, S., and Hong, S.S. 2013. Synergistic anticancer activity of HS-173, a novel P13K inhibitor in combination with Sorafenib against pancreatic cancer cells. *Cancer Lett.* 331: 250–261.

4. Chan, C.P., But, P., and Ho, J.W. 2002. Induction of rcl, a novel growth-related gene by Coptidis Rhizoma in rat H4IIE cells. *Life Sci.* 70: 1691–1699.

5. Ho, J.W., Leung, Y.K., and Chan, C.P. 2002. Herbal medicine in the treatment of cancer. *Curr. Med. Chem.* 2: 209.

6. Liu, J., Zhang, Q., Chen, K., Liu, J., Kuang, S., Chen, W., and Yu, Q. 2012. Small-molecule STAT3 signaling pathway modulators from *Polygonum cuspidatum*. *Planta Med.* 78: 1568–1570.

7. Wahl, O., Oswald, M., Tretzel, L., Herres, E., Arend, J., and Efferth, T. 2011. Inhibition of tumor angiogenesis by antibodies, synthetic small molecules and natural products. *Curr. Med. Chem.* 18: 3136–3155.

8. Lee, P.N.H. and Ho, W.S. 2013. Antiproliferative activity of gambogic acid isolated from *Garcinia hanburyi* in Hep3B and Huh7 cancer cells. *Oncol. Rep.* 29: 1744–1750.

9. Li, H., Zhu, H., Xu, C., and Yuan, J. 1998. Cleavage of Bid by caspase 8 mediates the mitochondrial damage in the Fas pathway of apoptosis. *Cell* 94: 491–501.

10. Ling, H., Yang, H., Tan, S.H., Chui, W.K., and Chew, E.H. 2010. 6-Shogaol, an active constituent of ginger, inhibits breast cancer cell invasion by reducing matrix metalloproteinase-9 expression via blockade of nuclear factor-κB activation. *Br. J. Pharmacol.* 161: 1763–1777.

11. Liu, Q., Shu, X., Sun, A., Sun, Q., Zhang, C., An, H., Liu, J., and Cao, X. 2008. Plant-derived small molecule albaconol suppresses LPS-triggered proinflammatory cytokine production and antigen presentation of dendritic cells by impairing NF-Kappa B activation. *Int. Immunopharmacol.* 8: 1103–1111.

12. Chiruvella, K.K. and Raghavan, S.C. 2011. A natural compound, methyl angolensate, induces mitochondrial pathway of apoptosis in Daudi cells. *Invest. New Drugs* 29: 583–592.

13. Chemler, S.R. 2009. Phenanthroindolizidines and phenanthroquinolizidines: Promising alkaloids for anti-cancer therapy. *Curr. Bioact. Compd.* 5: 2–19.

14. Lisowski, V., Léonce, S., Kraus-Berthier, L., Sopková-de Oliveira Santos, J., Pierré, A., Atassi, G., Caignard, D.H., Renard, P., and Rault, S. 2004. Design, synthesis, and evaluation of novel thienopyrrolizinones as antitubulin agents. *J. Med. Chem.* 47: 1448–1464.

15. Lesnard, A., Rault, S., and Dallemagne, P. 2013. MR22388, a novel anti-cancer agent with a strong FLT-3 ITD Kinase affinity. *Cancer Lett.* 331: 92–98.

16. Radhakrishnan, P., Bryant, V.C., Blowers, E.C., Rajule, R.N., Gautam, N., Anwar, M.M., Mohr, A.M., Grandgenett, P.M., Bunt, S.K., Arnst, J.L., Lele, S.M., Alnouti, Y., Hollingsworth, M.A., and Natarajan, A. 2013. Targeting the NF-Kappa B and mTOR pathways with a quinoxaline urea analog that inhibits IKKβ for pancreas cancer therapy. *Clin. Cancer Res.* 19: 2025–2035.

17. Liu, L.F., Desai, S.D., Li, T.K., Mao, Y., Sun, M., and Sim, S.P. 2000. Mechanism of action of camptothecin. *Ann. NY Acad. Sci.* 922: 1–10.

18. Liu, Y.P., Chen, H.L., Tzeng, C.C., Lu, P.J., Lo, C.W., Lee, Y.C., Tseng, C.H., Chen, Y.L., and Yang, C.N. 2013. TCH-1030 targeting on topoisomerase I induces S-phase arrest, DNA fragmentation, and cell death of breast cancer cells. *Breast Cancer Res. Treat.* 138: 383–393.

19. Goel, A., Kunnumakkara, A.B., and Aggarwal, B.B. 2008. Curcumin as 'Curecumin': From kitchen to clinic. *Biochem. Pharmacol.* 75: 787–809.

20. Thakkar, A., Sutaria, D., Grandhi, B.K., Wang, J., and Prabhu, S. 2013. The molecular mechanism of action of aspirin, curcumin and sulforaphane combinations in the chemoprevention of pancreatic cancer. *Oncol. Rep.* 29: 1671–1677.

21. Sutapa, M., Alok, C.B., Shirish, S., Abhishek, T., Syed, A.H., and Bhudev, C.D. 2011. Berberine modulates AP-1 activity to suppress HPV transcription and downstream signaling to induce growth arrest and apoptosis in cervical cancer cells. *Mol. Cancer* 10: 39–52.

22. Luo, Y., Hao, Y., Shi, T.P., Deng, W.W., and Li, N. 2008. Berberine inhibits cyclin D1 expression via suppressed binding of AP-1 transcription factors to CCND1 AP-1 motif. *Acta Pharmacol. Sin.* 29: 628–633.

23. Wu, H.L., Hsu, C.Y., Liu, W.H., and Yung, B.Y. 1999. Berberine-induced apoptosis of human leukemia HL-60 cells is associated with down-regulation of nucleophosmin/B23 and telomerase activity. *Int. J. Cancer* 81: 923–929.

24. Yip, N.K. and Ho, W.S. 2013. Berberine induces apoptosis via the mitochondrial pathway in liver cancer cells. *Oncol. Rep.* 30(3): 1107–1112.
25. Li, Y., Dowbenko, D., and Lasky, L.A. 2002. AKT/PKB phosphorylation of p21Cip/WAF1 enhances protein stability of p21Cip/WAF1 and promotes cell survival. *J. Biol. Chem.* 277(13): 11352–11361.
26. Reed, J.C. 1997. Bcl-2 family proteins: Regulators of apoptosis and chemoresistance in hematologic malignancies. *Semin. Hematol.* 34(4 Suppl 5): 9–19.
27. Boulares, A.H., Yakovlev, A.G., Ivanova, V., Stoica, B.A., Wang, G., Iyer, S., and Smulson, M. 1999. Role of poly(ADP-ribose) polymerase (PARP) cleavage in apoptosis. *J. Biol. Chem.* 274(33): 22932–22940.
28. Degterev, A. and Yuan, J. 2008. Expansion and evolution of cell death programmes. *Nat. Rev. Mol. Cell Biol.* 9(5): 378–390.
29. Fuchs, Y. and Steller, H. 2011. Programmed cell death in animal development and disease. *Cell* 147(4): 742–758.
30. Indran, I.R., Tufo, G., Pervaiz, S., and Brenner, C. 2011. Recent advances in apoptosis, mitochondria and drug resistance in cancer cells. *Biochim. Biophys. Acta* 1807(6): 735–745.
31. Kaufmann, T., Strasser, A., and Jost, P.J. 2012. Fas death receptor signalling: Roles of Bid and XIAP. *Cell Death Differ.* 19(1): 42–50.
32. Kurokawa, M. and Kornbluth, S. 2009. Caspases and kinases in a death grip. *Cell* 138(5): 838–854.
33. Pradelli, L.A., Bénéteau, M., and Ricci, J.E. 2010. Mitochondrial control of caspase-dependent and -independent cell death. *Cell. Mol. Life Sci.* 67(10): 1589–1597.
34. Van Herreweghe, F., Festjens, N., Declercq, W., and Vandenabeele, P. 2010. Tumor necrosis factor-mediated cell death: To break or to burst, that's the question. *Cell. Mol. Life Sci.* 67(10): 1567–1579.
35. Brumatti, G., Salmanidis, M., and Ekert, P.G. 2010. Crossing paths: Interactions between the cell death machinery and growth factor survival signals. *Cell. Mol. Life Sci.* 67(10): 1619–1630.
36. Fulda, S. and Vucic, D. 2012. Targeting IAP proteins for therapeutic intervention in cancer. *Nat. Rev. Drug Discov.* 11(2): 109–124.
37. Lopez, J. and Meier, P. 2010. To fight or die - Inhibitor of apoptosis proteins at the crossroad of innate immunity and death. *Curr. Opin. Cell Biol.* 22(6): 872–881.
38. Miura, M., Zhu, H., Rotello, R., Hartwieg, E.A., and Yuan, J. 1993. Induction of apoptosis in fibroblasts by IL-1 beta-converting enzyme, a mammalian homolog of the *C. elegans* cell death gene ced-3. *Cell* 75: 653–660.
39. Ola, M.S., Nawaz, M., and Ahsan, H. 2011. Role of Bcl-2 family proteins and caspases in the regulation of apoptosis. *Mol. Cell. Biochem.* 351(1–2): 41–58.
40. Lindsay, J., Esposti, M.D., and Gilmore, A.P. 2011. Bcl-2 proteins and mitochondria—Specificity in membrane targeting for death. *Biochim. Biophys. Acta* 1813(4): 532–539.
41. Speidel, D. 2010. Transcription-independent p53 apoptosis: An alternative route to death. *Trends Cell Biol.* 20(1): 14–24.

42. Rong, Y. and Distelhorst, C.W. 2008. Bcl-2 protein family members: Versatile regulators of calcium signaling in cell survival and apoptosis. *Annu. Rev. Physiol.* 70: 73–91.

43. Kantari, C. and Walczak, H. 2011. Caspase-8 and bid: Caught in the act between death receptors and mitochondria. *Biochim. Biophys. Acta* 1813(4): 558–563.

44. Lavrik, I.N. and Krammer, P.H. 2012. Regulation of CD95/Fas signaling at the DISC. *Cell Death Differ.* 19(1): 36–41.

45. Wajant, H. and Scheurich, P. 2011. TNFR1-induced activation of the classical NF-κB pathway. *FEBS J.* 278(6): 862–876.

46. Besson, A., Dowdy, S.F., and Roberts, J.M. 2008. CDK inhibitors: Cell cycle regulators and beyond. *Dev. Cell* 14(2): 159–169.

47. Yang, J.Y. and Hung, M.C. 2009. A new fork for clinical application: Targeting forkhead transcription factors in cancer. *Clin. Cancer Res.* 15(3): 752–757.

48. Malumbres, M. and Barbacid, M. 2009. Cell cycle, CDKs and cancer: A changing paradigm. *Nat. Rev. Cancer* 9(3): 153–166.

49. Musgrove, E.A., Caldon, C.E., Barraclough, J., Stone, A., and Sutherland, R.L. 2011. Cyclin D as a therapeutic target in cancer. *Nat. Rev. Cancer* 11(8): 558–572.

7 Inhibition of Cancer Growth by Herbal Medicines

Advanced cancer is a complex disease that demands treatment targeting critical organs and multiple signaling pathways. A traditional Chinese herbal formula is consisted of specific types of herbs with different pharmacological activities. Specific herbal composition that can be prescribed by experienced herbalists exerts multiple pharmacological effects that may target multiple regulatory pathways in cancerous tissues and organs. The Chinese herbal formulation known as Tien-Hsien Liquid (THL) can induce apoptosis in various cancer cells and can exert modulating activity to the immune system [1]. THL showed inhibitory effects on the migration and invasion activity of different cancer cells. It decreased the expression of MMP-2, MMP-9, and µPA and inhibited ERK1/2 activity. They showed multiarrays of activities including suppression of pulmonary metastasis of CT-26 cancer cells in mice, and inhibited the migration and tube formation of endothelial cells, and reduced the expression of MMP-2 and µPA in endothelial cells and suppressed neovascularization in mice. Besides its inhibitory effects on endothelial cells, THL attenuated hypoxia-induced HIF-1 alpha and vascular endothelial growth factor expression in vitro. In addition, THL inhibited the growth of human MDA-MB-231 breast cancer cells in xenografted mice. The study suggests that the specific traditional Chinese herbal composition should have broad spectrum of anticancer activities with pharmacological effects on different vital organs and immune response for treatment of cancers to produce health benefits. It is believed that mobilization of immune system with herbal medicines may play an important role in treatment of cancer and metastasis.

CYTOKINE AND INTERFERON-MODULATING PROPERTIES

Echinacea are commonly used for treatment of ailments including inflammation and infections. The pharmacological activities of *Echinacea* tinctures were studied using human peripheral blood mononuclear cells. Tinctures from seven species were examined for their immune-modulatory properties [2]. Table 7.1 shows the species of *Echinacea* that were evaluated in in vitro and in vivo studies.

TABLE 7.1

***Echinacea* spp. with
Immunomodulatory Properties**

1. *Echinacea angustifolia*
2. *Echinacea pallida*
3. *Echinacea paradoxa*
4. *Echinacea simulata*
5. *Echinacea tennesseensis*
6. *Echinacea sanguinea*
7. *Echinacea purpurea*

The immune-modulating properties of *Echinacea* spp. were compared by measuring the production of IL-2, IL-10, and interferon-gamma in vitro. Tinctures from *E. angustifolia*, *E. pallida*, *E. paradoxa*, and *E. tennesseensis* could stimulate proliferation and ameliorate IL-10 level, while *E. sanguinea* and *E. simulata* stimulated proliferation cytokine only. *E. purpurea* stimulated only IL-10. However, some of the extracts induced the production of influenza-specific IL-10 or IFN-gamma. Nevertheless, the immune response can vary depending on the type of stimuli.

AYURVEDIC PHARMACOLOGIC ACTIVITIES

Tinospora cordifolia Hook. f. and Thoms. are medicinal plants commonly used in Indian folk medicine for treatment of diseases including diabetes, pyrexia, spasm, inflammation, and hepatoprotection. *T. cordifolia* exhibits a broad spectrum of pharmacologic properties. It shows remarkable immunomodulatory activities [3]. The studies suggest that the immune response plays a pivotal role in modulating the functions of vital organs of humans. The amelioration of immune system can produce health benefits and is important in treatment of human diseases.

CYNOMORIUM WITH IMMUNOMODULATING PROPERTIES

Species of the genus Cynomoriaceae including *Cynomorium songaricum* Rupr. and *Cynomorium coccineum* L. are widely used for treatment various ous ailments including impotence, kidney disorder, and stomach ulcers in Asia. The herbal extracts of Cynomoriaceae are commonly used in health foods and cosmetics [4]. Chemical analysis of the active fractions of Cynomoriaceae extracts revealed that the active fractions contained flavonoids, terpenoids, phloroglucinol, and phenylpropanoids, which are the common phytochemicals found in herbal medicines. However, the composition of these phytochemicals is different among different species of the genus Cynomoriaceae. Like many other medicinal herbs, their crude extracts exhibited a wide spectrum of in vitro and in vivo pharmacological activities including modulation of immune system, which is important in any treatment of ailments. Amelioration of immune systems is believed to help balance "yin" and "yang" of the human body, and consequently, together with other medicinal herbs, ailments including cancer can be treated. Cynomoriaceae has aroused great interest and has become an important source of herbal medicines for treatment of various diseases. The potential health benefits of *Cynomorium* species have been reported. The studies provided supporting evidence for the clinical trial of these species in treating various diseases.

Although Cynomoriaceae has been used as folk medicine in Asia, the detail of pharmacological activities of the bioactive phytochemicals of Cynomoriaceae are lacking. Only a few of the active constituents have been reported to exhibit beneficial health benefits [5,6]. The *C. songaricum* and *C. coccineum* extracts are reported to show the immune-modulating effects. The chemical compositions of Cynomoriaceae extracts are worth investigated. The active principles and their pharmacological mechanism of actions and the molecular targets of these active phytochemicals should be characterized.

MYRTUS COMMUNIS LINN MODULATES T-CELL PROLIFERATION

Chemical analysis of phytochemical constituents of *M. communis* Linn identified myrtucommuacetalone and a novel phloroglucinol compound, myrtucommulone [7]. Together with other four polyphenolic compounds, myrtucommuacetalone and myrtucommulone can modulate the immune response. Myrtucommuacetalone exhibited significant inhibitory effects on nitric oxide synthase, and it exerted remarkable antiproliferative activity with $IC_{50} < 0.5$ µg/mL against T-cell proliferation. The other constituent, myricetin exerted a significant inhibition on zymosan-stimulated whole blood phagocytes reactive oxygen species (ROS) generation. Among the six constituents, only myrtucommulone and myricetin were active against the generation of phorbol 12-myristate 13-acetate (PMA)-stimulated ROS. PMA is a diester of phorbol and a potent tumor promoter often employed in biomedical research to activate the signal transduction enzyme protein kinase C (PKC). The studies indicated that an effective herbal phytochemical should act against different targets and multiple pathways.

MODULATION OF SPLENOCYTE FUNCTIONS BY *CYPERUS ROTUNDUS*

Cyperaceae is folkloric medicine used to treat stomach ailments and inflammatory diseases. *C. rotundus* Linn is the species of the genus Cyperaceae. It shows anti-inflammatory and antigenotoxic activities. The antioxidant activity and the modulation of splenocyte functions by *C. rotundus* Linn extracts were demonstrated in mice [8]. The active fraction of *C. rotundus* reduced the ear edema that was induced by xylene and decreased the number of abdominal contractions caused by acetic acid in mice. The active fraction of *C. rotundus* could significantly enhance lymphocyte proliferation at 1 mg/mL.

PHARMACOLOGICAL ACTIVITY OF LYCOPENE

A nutritionally relevant phytochemical compound such as lycopene found in fruits and other herbs is believed to induce health-promoting effects by modulating endocrine and immune systems, metabolic, and signaling pathways [9]. The extracts prepared from tomato and grape seeds showed a peculiar composition with high lycopene concentration and other carotenoids and polyphenols. The extracts prepared by supercritical carbon dioxide showed a higher in vitro antioxidant activity compared with synthetic lycopene and beta-carotene, suggesting a pure chemical compound may not produce the pharmacological activity. The studies imply that a natural product can function more effectively as a conjugated complex. When the single compound was isolated and purified from the active fraction, it may not exhibit the pharmacological activity. The active fraction is therefore more effective and should be used for treatment of ailments. The tomato extract was showed to enhance the gap junction intercellular communication and increased CX43 expression in keratinocytes. The lycopene-containing extract can completely reverse the gap junction intercellular communication induced by 10 mm mercuric chloride. The results suggest that the lycopene can activate the action potential through antioxidative activity.

PLANT POLYPHENOLS IN THE TREATMENT OF CANCER

Epidemiological studies indicated that food supplements and plant-derived foods have low incidence rates of cancers. Recent reports [10–13] suggest a variety of phytochemicals, including flavonoids and polyphenols, are believed to be associated with the anticancer activity. Some of these flavonoids and polyphenols that are present in foods exhibit anticancer properties. Polyphenolic compounds derived from herbal medicines have been

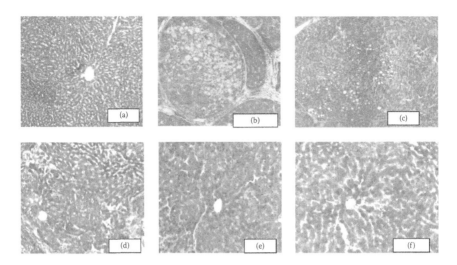

FIGURE 7.1   (See color insert.) Anticancer activities of sinigrin in liver cancer in rats. (a) Normal, (b) untreated, (c) DOX, (d) SIN 10, (e) SIN 15, and (f) SIN 25.

shown to attenuate the growth and metastasis of cancers in cells in vitro and in animals. Figure 7.1 shows the anticancer activities of sinigrin in liver cancer [10]. The molecular mechanisms that polyphenolics modulate cellular functions are complex and involve regulation of growth factor–receptor interactions and other signaling cascades, including cell cycle and transcription regulation. A major focus has been the inhibitory effects of polyphenols on the stress-induced NF-kappa B and AP-1 signal cascades in cancer cells. These are considered as the major molecular targets for drug development [11]. It is believed that multimolecular targets should be modulated by combination therapy with phytochemicals. A single phytochemical may not be effective enough to provide curative measure or palliative care of cancers.

IMMUNOMODULATION OF GINSENG ROOT IN CHEMOPREVENTION

Ginseng was known to have immunomodulatory effects, and it is commonly used in combination with other herbal formulations. However, the pharmacological activity of ginseng remains unclear. The aqueous extracts of ginseng showed upregulation of the production of nitric oxide and tumor necrosis factor-alpha (TNF-alpha) and interleukin-6, while the ethanolic extracts did not show the activities [12]. The ethanol extract suppressed lipopolysaccharide-induced macrophage nitric oxide and TNF-alpha production. Besides, the macrophage-stimulating activity of the aqueous extract was inhibited by the ethanolic extract. Chemical analysis of the aqueous fraction showed the bioactive constituents contained polysaccharide, while the ethanolic fraction did not contain polysaccharide. The polysaccharide in the aqueous fraction is believed to produce the immunostimulatory activities in humans.

IMMUNOMODULATION ACTIVITY OF *LORANTHUS MICRANTHUS* LINN

Loranthus micranthus is the folk medicinal plant in Nigeria. It was shown to exhibit antidiabetic, antimicrobial, and antihypertensive activities. Other study showed that *L. micranthus* possesses immunomodulating activities [13]. Phytochemical analysis of the herbal extracts showed the presence of flavonoids, polyphenols, terpenoids, tannins, saponins, and carbohydrates. The active fractions of an herbal medicine commonly contain most of these phytochemicals in different compositions that characterize the pharmacological properties of herbal medicines. Detailed investigation on the active fractions and the constituents are important to characterize the pharmacological activities of active fractions.

Alkylamides prepared from genus *Echinacea* showed inductive effects on immune responses in the presence of endotoxins [14]. The immune-modulating activities of *Echinacea* and some of active phytochemical constituents exerted significant effects on basal NF kappa B expression in the presence of endotoxin including lipopolysaccharide. NF kappa B expression was decreased in Jurkat cells, a human T-cell line, in the presence of endotoxin but was reversed by cichoric acid, and the alkylamides prepared from *Echinacea* root extracts. Both *E. angustifolia* and *E. purpurea* contained cichoric acid and alkylamides. The studies indicated that two basic forms of alkylamide present in *Echinacea* may have opposite effects on the NF kappa B expression. The stimulatory and suppressive activities of alkylamides suggest the phytochemical composition of an herbal preparation could change the immune-modulating activities of *Echinacea* alkamides.

Among the active constituents, polysaccharides appear to play a role in immunomodulatory activity. Carbohydrates and other phytochemical compounds in the active fraction of *Pleurotus* fruiting bodies showed lymphoproliferative and stimulating response [15]. The immunomodulating properties of *Pleurotus* powder is attributed to the immunostimulating activities of *Pleurotus* fruiting bodies.

Herbs from the genus *Hedysarum* were reported to show immunomodulating activities. Table 7.2 shows the common species with medicinal properties in genus *Hedysarum*. They have been used in folk medicine in China for hundreds of years. There are 155 compounds including flavonoids,

TABLE 7.2

Species of *Hedysarum* with Immunomodulating Activities

Genus *Hedysarum*

1. *Hedysarum polybotrys* Hand
2. *Hedysarum limprichtii* Hlbr
3. *Hedysarum vicioider* Turcz
4. *Hedysarum smithianum*

triterpenes, coumarins, alkaloids, carbohydrates, and benzofuran found in genus *Hedysarum* [16].

The phytochemical constituents of *Hedysarum* exhibit a wide spectrum of medical properties including antitumor, antidiabetic, antiaging, antihypertensive, and anti-inflammatory properties, yet the immune response may vary based on stimulus. *Hedysarum* species have emerged as a source of an alternative medicine. Many studies have provided evidence for the therapeutic efficacy of these species in treating various conditions and possible mechanisms were reported. However, more research is required for the development of new drugs and therapy strategies for the treatment of various diseases, especially cancer and diabetes. Therefore, the ethnopharmacology and phytochemistry of *Hedysarum* species would provide useful data for exploitation of the species.

ANTI-INFLAMMATION EFFECT OF HERBAL MEDICINE

The anti-inflammatory activity of methanolic and petroleum ether extracts of herbal medicine was reported to show multiarrays of medicinal properties [17]. A time-dependent evaluation of anti-inflammatory activity suggested that the extract shows activity at various acute phases of inflammation. This anti-inflammatory property might be attributed by the β-sitosterol content of herbs with anti-inflammatory constituents [18]. The results on herbal extracts showed the anti-inflammatory effect is evolved in early phase of inflammation [18].

PHYTOCHEMICAL PROPERTIES OF HERBAL MEDICINE

The phytochemicals are the major component responsible for the pharmacological and medicinal properties of a plant material [19]. A long history of traditional medicinal usage indicated the presence of bioactive alkaloids including polyphenols and flavonoids, but the pharmacological effects of the majority of the principles remain unknown.

DIFFERENT PARTS OF A MEDICINAL HERB CONTAIN VARIOUS TYPES OF PHYTOCHEMICALS

Legler [20] reported the presence of L-rhamnose, D-fucose, D-chinovose (6-deoxy-D-glucose), D-glucose, convolvulinolate (11-hydroxy-pentadecane acid), jalapinolate (11-hydroxypalmitic acid), 7-hydroxy-decane acid, and ipurolic acid (3,11-dihydroxy-tetradecane acid) in the latex of *Ipomoea carnea* and subspecies *I. fistulosa*. Tirkey et al. [21] reported the phytochemical studies of the leaves of *I. carnea* and found the presence of alkaloids, reducing sugars, glycosides, and tannins in the dried powdered leaves, while acacetin-7-galactoside, flavone glycoside, and saponin of unknown chemical structure called ipomotocin are contained in the latex. In recent studies, the leaves, stem, and flowers of *I. carnea* have been reported to contain appraisable amount of polyphenols (30–70 mg catechol equivalent/g dry material) and flavonoid (80–120 mg quercetin equivalent/g dry material) [22]. Arora et al. [23] reported the presence of alkaloids, carbohydrates, tannins, phenolic compounds, proteins and amino acid, terpenoids and sterols, and saponins in methanol extract of leaves and flowers of *I. carnea*. Among these phytochemicals, only alkaloids of polyhydroxylated class and terpenoid in the leaves, flowers, and seeds have been characterized partially. Chromatographic study on the leaves, flowers, and seeds resulted in the isolation of swainsonine, 2-epi-lentiginosine, calystegines B1, calystegines B2, calystegines B3, calystegines C1, and N-methyl-trans-4-hydroxy-L-proline at varying combination and concentration [24]. A gas chromatography–mass spectrometry study on the hexane extract of *I. carnea* showed the presence of a panel of 13 compounds including hexadecanoic acid, stearic acid, 1,2-diethyl phthalate, n-octadecanol, octacosane, hexatriacontane, tetracontane, and 3-diethylamino-1-propanol [25]. In case of protein, only one protein molecule has been isolated and characterized from the latex of *I. carnea*, known as "carnein," which is a serine protease with a molecular weight of 80.24 kDa [26].

There has been the resurgence of interests on phytomedicine in recent decade. Modern medical science is looking for novel drug development through in part the establishment of traditional medicine, especially phytomedicine, in parallel with modern chemical medicine. In fact, the phytomedicines are the source of many chemical medicines, too. Phytotherapy is believed to be one of the mainstream therapeutic approaches in future, for its safety and potency. *I. carnea* has been used for thousands of years, but detail of its medicinal properties is lacking. Most of the researches on the medicinal issue of *I. carnea* have explored only in in vitro cellular and animal models. To establish the promising medicinal plant for human drug development, more preclinical and clinical studies are required. The basic research on the therapeutic effect of herbal medicine such as *I. carnea* and its phytoconstituents using disease-specific animal (e.g., diabetic, immune deficient, and cancerous) model first provides a preclinical scientific basis of the usage of a specific herb medicine. In vitro animal cell experiments (e.g., mouse embryonic fibroblast 3T3, cerebral cortex b, embryo NIH-3T3, pancreas RIN-5F, myocardial endothelial MyEnd, and hamster kidney BHK-21 cell lines) may provide a feasible technique to figure out both organ-specific pharmacological effect and active principles of *I. carnea* besides preliminary toxicological evaluation. A stepwise evaluation of the therapeutic effect of *I. carnea* using normal and diseased human cell line can provide a strong scientific basis of the *I. carnea* usage. Finally, herbal medicine and/or its phytoconstituents of clinical significance might be applied clinically only after safety evaluation. More scientific studies (both clinical and basic studies) can give an herbal medicine such as *I. carnea* a notable position in modern phytotherapy research.

RESVERATROL INHIBITS COLON CANCER CELL PROLIFERATION

Colorectal cancer is one of the common malignancies in humans. The incidence of colon cancer increases with younger age [27]. The current treatments of colon cancer showed limitations [28]. Treatments of cancer with natural products have become an alternative approach [28,29]. Many herbs or their active compounds have been studied and applied preclinically. Anticancer drug development including anthocyanidins, genistein, and quercetin are some of the good examples [30,31]. Grape extract was shown to have therapeutic effects on cardiovascular disease and cancer [32,33]. Polyphenolic compounds are present in grapes. Resveratrol (Figure 7.2), the active polyphenol, is also present in beans and grapes [34]. Resveratrol was reported to inhibit obesity and promote apoptosis in various tumor cells including colon, breast, and prostate cancer cells [35–37].

PI3K/Akt signaling plays a critical role in modulating cell viability and programmed cell death [38]. The PI3K/Akt was regulated by phosphatase and tensin homolog (PTEN) [39]. Wnt/β-catenin signaling pathway is upregulated by phosphorylation of glycogen synthase 3 kinase β (GSK3β), an important negative regulator for Wnt/β-catenin signaling

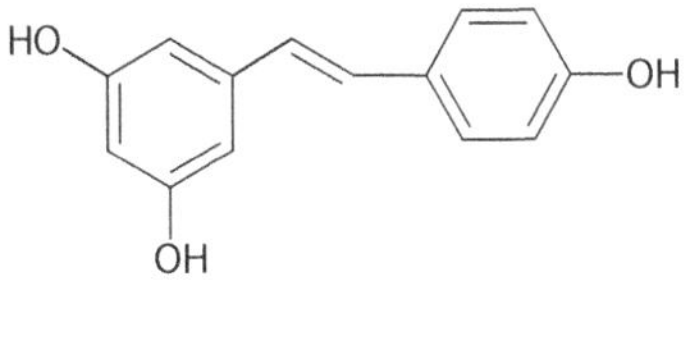

FIGURE 7.2   Chemical structure of resveratrol.

pathway, when PI3K/Akt signal is activated [40]. Resveratrol was found to exhibit prominent antiproliferation and induce apoptosis of HCT116 cells. Resveratrol upregulated the expression of PTEN and inhibited the activation of PI3K/Akt signaling pathway and to inhibit the Wnt/β-catenin signaling transduction.

Although the treatment for colon cancer has advanced substantially, the prognosis remains unsatisfactory [41]. New anticancer agents are warranted. Resveratrol has potent antiproliferation activity in human colon cancer cells by inhibiting PI3K/Akt signaling through upregulating the expression of PTEN and blocking Wnt/β-catenin signaling transduction, respectively [42]. Several studies have proved that resveratrol can inhibit proliferation and induce apoptosis of various cancer cells including colon cancer cells [32,43,44]. These data confirmed that resveratrol has the potential to be an anticancer agent. Resveratrol could induce apoptosis through the suppression of Wnt pathway and activation of p53 signaling pathways in human colon cancer [45]. The expression of MicroRNA-21 participates in the inhibition of prostate cancer growth and metastasis initialized by resveratrol [46]. p38 and PI3K signaling pathways are reported to be involved in the anticancer activity of resveratrol [47]. Wnt, PI3K/Akt, and p38 signaling pathways are all essential for cell proliferation and differentiation and found to be aberrant in many cancers.

PTEN, a tumor suppressor often mutated or lost in many cancers, acts as a phosphatase to specifically catalyze PIP3 dephosphorylation at the 3-phosphate of the inositol ring and turn PIP3 to PIP2 through which it negatively regulates PI3K/Akt signaling. PI3K/Akt signaling has been associated with many cellular functions, including proliferation, differentiation, motility, and survival. PTEN may play an important role in early stages of sporadic colorectal carcinogenesis [48]. However, exogenous expression of PTEN can potentiate the proliferation inhibitory effect of resveratrol in these cells. These data indicate that PTEN may be important for the antiproliferation effect of resveratrol in HCT116 cells. Further analysis confirmed that the upregulated PTEN by resveratrol is correlated with the inactivation of PI3K/Akt signaling by decreasing the phosphorylation of Akt1/2 in HCT116 cells. The results suggest that PTEN/PI3K/Akt is involved in the proliferation inhibitory effect of Res in HCT116 cells.

Wnt signaling, including canonical and noncanonical Wnt signaling, plays an important role in embryogenesis and development. The β-catenin plays a pivotal role in canonical Wnt signaling.

The activation of PI3K/Akt can activate the canonical Wnt signaling through the phosphorylation of GSK-3β by the phosphorylated Akt1/2, blocking the formation of β-catenin destroying complex [40]. Hence, the upregulation of PTEN may inhibit the canonical Wnt signaling by promoting the degradation of β-catenin. Resveratrol can decrease the phosphorylation of GSK-3β, the protein level of β-catenin, and the β-catenin/Tcf4 reporter activity, but it may not result from the decreased phosphorylation of GSK-3β by PTEN/PI3K/Akt signaling. Resveratrol can inhibit the Wnt/β-catenin signaling transduction, but it may not result from the upregulation of PTEN in HCT116 cells.

Resveratrol was shown to inhibit the proliferation and promote apoptosis in colon cancer cells. These activities of resveratrol may be mediated by PI3K/Akt signaling through upregulating the expression of PTEN and reducing the Wnt/β-catenin signaling transduction through inhibiting the expression of β-catenin, respectively.

ANTICANCER ACTIVITIES OF REACTIVE OXYGEN SPECIES

Growing evidence indicates that ROS play a critical role in many disorders of the cardiovascular system [49–51]. The accumulation of ROS in mitochondria can lead to apoptotic cell death and ROS may also have direct effects on cellular structure and function, including myocardial remodeling and failure [46]. Accordingly, inhibiting ROS production or the enhancement of ROS scavenging could be used for treating these cardiovascular disorders. Antioxidant compounds may be useful for ROS-related diseases, which have resulted in increasing demand for alternative and safe antioxidants isolated from natural sources such as plants and herbal medicine [40]. Hispidin has already been proven to exert potent free radical scavenging effects via its antioxidant properties, which might be used to eliminate the key mediators for the oxidative stress [49,52,53]. The results demonstrate a beneficial effect of hispidin on the myocardial response to oxidative stress. In vitro studies suggest a potential mechanism of this helpful effect in which hispidin attenuates H_2O_2-induced apoptosis in H9c2 cardiomyoblast cells by upregulating specific survival proteins such as Akt/GSK-3β and ERK1/2, and scavenging ROS generation via the activation of heme oxygenase-1 (HO-1) and catalase (CAT).

The chemical structure of hispidin, 6-(3,4-dihydroxystyryl)-4-hydroxy-2-pyrone, is very similar to the structure of strong antioxidants such as resveratrol derivatives found in a number of plant species [54]. The strong radical scavenging activities of resveratrol-like stilbene found in grapes [55]. It has been reported that the catechol moiety of hispidin may contribute to its antioxidative activity [56]. The study showed hispidin pretreatment reduced ROS production and increased HO-1 and CAT expression in H_2O_2-treated H9c2 cells using H_2O_2 to generate intracellular ROS. These results suggest that the protective effects of hispidin are due to its ROS scavenging ability. Several enzymes convert ROS into compounds that are either harmless or rapidly metabolized. These enzymes prominently include SODs that convert superoxide into oxygen and hydrogen peroxide such as CAT, which converts H_2O_2 into O_2 and water, thereby aiding in the maintenance of cell membrane integrity and preventing the activation of cellular signaling processes that may lead to diseases such as oxidative cardiovascular injury [57–59]. The results indicated that hispidin exerts its protective effect in part by enhancing antioxidant enzyme activity, thereby attenuating the oxidative damage.

A previous study indicated that hispidin pretreatment increased the expression of Bcl-2, as it decreased the expression of proapoptotic Bax [49]. Therefore hispidin might exert protective effects by maintaining mitochondrial function and modulating the balance of antiapoptotic protein Bcl-2 and proapoptotic protein Bax; thus, active anticancer phytochemicals may act in similar action in modulating the release of proapoptotic molecules from the mitochondrial intermembrane space into the cytoplasm, resulting in programmed cell death [60].

In addition, a group of survival protein kinases that include Akt and ERK1/2 that confer powerful cytoprotection provides an amenable pharmacological target for cardioprotection [61]. The antiapoptotic mechanism of hispidin may be responsible for the activation of Akt and ERK1/2. Akt, a serine/threonine kinase, is activated subsequently to the production of PIP3 by phosphatidylinositol-3-kinase (PI3K), and mediates several functions through the phosphorylation and inactivation of the proapoptotic kinase, glycogen synthase kinase-3 (GSK-3α/β) [62].

So far, three major MAPK signaling pathways—ERK1/2, p38 MAPK, and JNK—are commonly known to be modulated by anticancer agents. They are activated in response to myocardial I/R. The activation of ERK1/2 (beneficial effect) and p38 MAPK–JNK (a deleterious effect) exerts opposite effects on myocardial apoptosis and cardiac function recovery [63]. The activation of either the PI3K/Akt or the ERK1/2 pathway inhibits the conformational change in Bax required for its translocation to the mitochondria, therefore preventing apoptosis [64,65]. Furthermore, the upregulation of ERK1/2 and Akt inactivates the caspase cascade, a proposition that is supported by inhibiting caspase-3 and caspase-9 activation [66,67].

Anticancer agents can modulate cardiomyocyte damage caused by oxidative stress through PKC inhibition [68]. Drugs that target PKC include Gö6983, a broad-spectrum PKC inhibitor, which restores cardiac function and attenuates the deleterious effects in isolated perfused rat heart after PMN-induced I/R injury [69]. It is plausible that anticancer agents may inhibit PKC activity by binding to its ATP binding sites, and hence attenuate cardiac contractile dysfunction, a process similar to the mechanism of action of other PKC inhibitors [70,71].

Oxidative stress can therefore be regulated by antioxidants, such as those active polyphenols and flavonoids from herbal medicine, through modulation of caspase-3, Bax, and Bcl-2, as well as by the activation of

Akt/GSK-3β and ERK1/2 signaling pathways, and by enhancing antioxidant enzyme systems.

Integrative medicine is a relatively new discipline that combines Western medicine with alternative or complementary medicine, to reap the benefits of both forms of medicine in treatment of cancer. While the evidence-based clinical research drives the management and treatment of diseases in conventional Western medicine, alternative or complementary medicine is based on traditional treatment with therapeutic effects, yet health benefits have been developed throughout history, dating back to the ancient cultures in the Middle East, Africa, and China. In spite of the lack of statistical and scientific results on most alternative medicine techniques, these methodologies have been practiced for centuries with great acceptance in many countries. Consequently, the controversy in the use of these alternative modes of therapy remains. It has been a hurdle for those trained in Western medicine to adopt the integrated approach to treatment of ailments. Although Western medicine often fails to adequately guarantee our well-being and good health, an increasing number of studies reveal the benefits of these alternative treatments. Integrative medicine may be a futile attempt to optimize the treatment strategy. Studies to evaluate the scientific analysis behind ancient medical techniques should be updated all the time. Of particular interest is the effect of traditional medicine, herbal formulations, and manipulative techniques on the immune system, and its application in the treatment of autoimmune and allergic diseases. Herbal plants or medicines may offer potential application in the treatment of allergic or autoimmune diseases.

REFERENCES

1. Chia, J.S., Du, J.L., Hsu, W.B., Sun, A., Chiang, C.P., and Wang, W.B. 2010. Inhibition of metastasis, angiogenesis, and tumor growth by Chinese herbal cocktail Tien-Hsien Liquid. *BMC Cancer* 10: 175.
2. McCann, D.A., Solco, A., Liu, Y., Macaluso, F., Murphy, P.A., Kohut, M.L., and Senchina, D.S. 2007. Cytokine- and interferon-modulating properties of *Echinacea* spp. root tinctures stored at –20 degrees C for 2 years. *J. Interferon Cytokine Res.* 27: 425.
3. Upadhyay, A.K., Kumar, K., Kumar, A., and Mishra, H.S. 2010. Tinospora cordifolia (Willd.) Hook. f. and Thoms. (Guduchi)—Validation of the ayurvedic pharmacology through experimental and clinical studies. *Int. J. Ayurveda Res.* 1: 112–121.
4. Cui, Z.H., Guo, Z.Q., Miao, J.H., Wang, Z.W., Li, Q.Q., Chai, X.Y., and Li, M.H. 2013. The genus *Cynomorium* in China: An ethnopharmacological and phytochemical review. *J. Ethnopharmacol.* 147: 1–15.
5. Yin, P. 2011. Health liquor of traditional Chinese medicines for preventing dizziness and Sonitus. *CN* 102274345A: 20111214.

6. Yu, F.R., Liu, Y., Cui, Y.Z., Chan, E.Q., Xie, M.R., McGuire, P.P., and Yu, F.H. 2010. Effects of a flavonoid extract from *Cynomorium songaricum* on the swimming endurance of rats. *Am. J. Chin. Med.* 38: 65–73.

7. Choudhary, M.I., Khan, N., Ahmad, M., Yousuf, S., Fun, H.K., Soomro, S., Asif, M., Mesaik, M.A., and Shaheen, F. 2013. New inhibitors of ROS generation and T-cell proliferation from *Myrtus communis. Org. Lett.* 15: 1862–1865.

8. Soumaya, K.J., Dhekra, M., Fadwa, C., Zied, G., Ilef, L., Kamel, G., and Leila, C.G. 2013. Pharmacological, antioxidant, genotoxic studies and modulation of rat splenocyte functions by *Cyperus rotundus* extracts. *BMC Compliment. Altern. Med.* 13: 28.

9. Leone, A., Zefferino, R., Longo, C., Leo, L., and Zacheo, G. 2010. Supercritical CO(2)-extracted tomato Oleoresins enhance gap junction intercellular communications and recover from mercury chloride inhibition in keratinocytes. *J. Agric. Food Chem.* 58: 4769–4778.

10. Meng, J., Cheung, W.M., Yu, V., Zhou, Y., Tong, P.H., and Ho, W.S.J. 2014. Anti-proliferative activities of sinigrin on carcinogen-induced hepatotoxicity in rats. *PLoS ONE* 9(10): e110145.

11. Wahle, K.W., Brown, I., Rotondo, D., and Heys, S.D. 2010. Plant phenolics in the prevention and treatment of cancer. *Adv. Exp. Med. Biol.* 698: 36–51.

12. Azike, C.G., Charpentier, P.A., Hou, J., Pei, H., and Lui, E.M.K. 2011. The Yin and Yang actions of North American ginseng root in modulating the immune function of macrophages. *Chin. Med.* 6: 21.

13. Omeje, E., Osadebe, P.O., Okoyea, F.B.C., Agwu, A.K., and Esimone, C.O. 2008. Immunomodulation and Nigerian mistletoe immunomodulatory activities of n-hexane and methanol extracts of *Loranthus micranthus* Linn. parasitic on *Parkia biglobosa. Asian Pac. J. Trop. Med.* 1: 48.

14. Matthias, A., Banbury, L., Bone, K.M., Leach, D.N., and Lehmann R.P. 2008. Echinacea alkylamides modulate induced immune responses in T-cells. *Fitoterapia* 79: 53–58.

15. Llaurado, G., Morris, H.J., Lebeque, Y., Gutierrez, A., Fontaine, R., Bermudez, R.C., and Perraud-Gaime, I. 2013. Phytochemical screening and effects on cell-mediated immune response of *Pleurotus* fruiting bodics powder. *Food Agric. Immunol.* 24: 295–304.

16. Dong, Y.M., Tang, D.Y., Zhang, N., Li, Y., Zhang, C., Li, L., and Li, M.H. 2013. Phytochemicals and biological studies of plants in genus *Hedysarum. Chem. Cent. J.* 7: 124.

17. Ruchi, J., Nilesh, J., and Surendar, J. 2009. Evaluation of anti-inflammatory activity of *Ipomoea fistulosa* linn. *Asian J. Pharm. Clin. Res.* 2: 64–67.

18. Khalid, M.S., Singh, R.K., Reddy, I.V., Kumar, S.J., Kumar, B.S., Kumar, G.N., and Rao, K.S. 2011. Anti-inflammatory activity of aqueous extract of *Ipomoea carnea* jacq. *Pharmacology* (Online) 1: 326–331.

19. Fatima, N., Rahman, M.M., Khan, M.A., and Fu, J. 2014. A review on *Ipomoea carnea*: Pharmacology, toxicology and phytochemistry. *J. Complement. Integr. Med.* 11(2): 55–62.

20. Legler, G. 1965. Die bestandteile des giftigen glykosidharzes aus *Ipomoea fistulosa* Mart. ex Choisy. *Phytochemistry* 4: 29–41.

21. Tirkey, K., Yadava, R.P., Mandal, T.K., and Banerjee, N.L. 1988. The pharmacology of *Ipomoea carnea*. *Indian Vet J.* 65: 206–210.
22. Khatiwora, E., Adsul, V.B., Kulkarni, M.M., Deshpande, N.R., and Kashalkar, R.V. 2010. Spectroscopic determination of total phenol and flavonoid contents of *Ipomoea carnea*. *Int. J. ChemTech Res.* 2: 1698–1701.
23. Arora, S., Kumar, D., and Shiba. 2013. Phytochemical, anti-microbial and anti-oxidant activities of methanol extract of leaves and flowers of *Ipomoea cairica*. *Int. J. Pharm. Pharm. Sci.* 1: 198–202.
24. Adsul, V., Khatiwora, E., Kulkarni, M., Tambe, A., Pawar, P., and Deshpande, N. 2009. GCMS study of fatty acids, esters, alcohols from the leaves of *Ipomoea carnea*. *Int. J. Pharmtech Res.* 1: 1224–1226.
25. Haraguchi, M., Gorniak, S.L., Ikeda, K., Minami, Y., Kato, A., Watson, A.A., Nash, R.J., Molyneux, R.J., and Asano, N. 2003. Alkaloidal components in the poisonous plants, *Ipomoea carnea* (Convolvulaceae). *J. Agric. Food Chem.* 51: 4995–5000.
26. Patel, A.K., Singh, V.K., and Jagannadham, M.V. 2007. Carnein, a serine protease from noxious plant weed *Ipomoea carnea* (Morning Glory). *J. Agric. Food Chem.* 55: 5809–5818.
27. Brunagel, G., Vietmeier, B.N., Bauer, A.J., Schoen, R.E., and Getzenberg, R.H. 2002. Identification of nuclear matrix protein alterations associated with human colon cancer. *Cancer Res.* 62: 2437–2442.
28. Cragg, G.M., Grothaus, P.G., and Newman, D.J. 2009. Impact of natural products on developing new anti-cancer agents. *Chem. Rev.* 109(7): 3012–3043.
29. Mishra, B.B. and Tiwari, V.K. 2011. Natural products: An evolving role in future drug discovery. *Eur. J. Med. Chem.* 46(10): 4769–4807.
30. Korkina, L.G., De Luca, C., Kostyuk, V.A., and Pastore, S. 2009. Plant polyphenols and tumors: From mechanisms to therapies, prevention, and protection against toxicity of anti-cancer treatments. *Curr. Med. Chem.* 16(30): 3943–3965.
31. Khan, N., Afaq, F., and Mukhtar, H. 2008. Cancer chemoprevention through dietary antioxidants: Progress and promise. *Antioxid. Redox Signal.* 10(3): 475–510.
32. Jang, M., Cai, L., Udeani, G.O., Slowing, K.V., Thomas, C.F., Beecher, C.W., Fong, H.H., Farnsworth, N.R., Kinghorn, A.D., Mehta, R.G., Moon, R.C., and Pezzuto, J.M. 1997. Cancer chemopreventive activity of resveratrol, a natural product derived from grapes. *Science* 10: 218–220.
33. Das, S., Santani, D.D., and Dhalla, N.S. 2007. Experimental evidence for the cardioprotective effects of red wine. *Exp. Clin. Cardiol.* 12(1): 5–10.
34. Gehm, B.D., McAndrews, J.M., Chien, P.Y., and Jameson, J.L. 1997. Resveratrol, a polyphenolic compound found in grapes and wine, is an agonist for the estrogen receptor. *Proc. Natl. Acad. Sci. U.S.A.* 94(25): 14138–14143.
35. Yang, H.L., Chen, W.Q., Cao, X., Worschech, A., Du, L.F., Fang, W.Y., Xu, Y.Y., Stroncek, D.F., Li, X., Wang, E., and Marincola, F.M. 2009. Caveolin-1 enhances resveratrol-mediated cytotoxicity and transport in a hepatocellular carcinoma model. *J. Trans. Med.* 25(7): 22–26.

36. Lu, R. and Serrero, G. 1999. Resveratrol, a natural product derived from grape, exhibits antiestrogenic activity and inhibits the growth of human breast cancer cells. *J. Cell. Physiol.* 179(3): 297–304.

37. Fouad, M.A., Agha, A.M., Merzabani, M.M., and Shouman, S.A. 2013. Resveratrol inhibits proliferation, angiogenesis and induces apoptosis in colon cancer cells: Calorie restriction is the force to the cytotoxicity. *Hum. Exp. Toxicol.* 32(10): 1067–1080.

38. Fresno Vara, J.A., Casado, E., de Castro, J., Cejas, P., Belda-Iniesta, C., and González-Barón, M. 2004. PI3K/Akt signalling pathway and cancer. *Cancer Treat. Rev.* 30(2): 193–204.

39. Takeda, K., Kanekura, T., and Kanzaki, T. 2004. Negative feedback regulation of phosphatidylinositol 3-kinase/Akt pathway by over-expressed cyclo-oxygenase-2 in human epidermal cancer cells. *J. Dermatol.* 31(7): 516–523.

40. Robertson, B.W. and Chellaiah, M.A. 2010. Osteopontin induces beta-catenin signaling through activation of Akt in prostate cancer cells. *Exp. Cell Res.* 316(1): 1–11.

41. Aggarwal, S. and Chu, E. 2005. Current therapies for advanced colorectal cancer. *Oncology* 19: 589–595.

42. Liu, Y.Z., Wu, K., Huang, J., Liu, Y., Wang, X., Meng, Z.J., Yuan, S.X., Wang, D.X., Luo, J.Y., Zuo, G.W., Yin, L.J., Chen, L., Deng, Z.L., Yang, J.Q., Sun, W.J., and He, B.C. 2014. The PTEN/PI3K/Akt and Wnt/β-catenin signaling pathways are involved in the inhibitory effect of resveratrol on human colon cancer cell proliferation. *Int. J. Oncol.* 45(1): 104–112.

43. Gulvady, A.A., Ciolino, H.P., Cabrera, R.M., and Jolly, C.A. 2013. Resveratrol inhibits the deleterious effects of diet-induced obesity on thymic function. *J. Nutr. Biochem.* 24: 1625–1633.

44. Fang, J.Y., Li, Z.H., Li, Q., Huang, W.S., Kang, L., and Wang, J.P. 2012. Resveratrol affects protein kinase C activity and promotes apoptosis in human colon carcinoma cells. *Asian Pac. J. Cancer Prev.* 13: 6017–6022.

45. Vanamala, J., Reddivari, L., Radhakrishnan, S., and Tarver, C. 2010. Resveratrol suppresses IGF-1 induced human colon cancer cell proliferation and elevates apoptosis via suppression of IGF-1R/Wnt and activation of p53 signaling pathways. *BMC Cancer* 10: 238.

46. Sheth, S., Jajoo, S., Kaur, T., Mukherjea, D., Sheehan, K., Rybak, L.P., and Ramkumar, V. 2012. Resveratrol reduces prostate cancer growth and metastasis by inhibiting the Akt/MicroRNA-21 pathway. *PLoS ONE* 7(12): e51655.

47. Gweon, E.J. and Kim, S.J. 2013. Resveratrol induces MMP-9 and cell migration via the p38 kinase and PI-3K pathways in HT1080 human fibrosarcoma cells. *Oncol. Rep.* 29: 826–834.

48. Waniczek, D., Śnietura, M., Młynarczyk-Liszka, J., Pigłowski, W., Kopeć, A., Lange, D., Rudzki, M., and Arendt, J. 2013. PTEN expression profiles in colorectal adenocarcinoma and its precancerous lesions. *Pol. J. Pathol.* 64: 15–20.

49. Kim, D.E., Kim, B., Shin, H.S., Kwon, H.J., and Park, E.S. 2014. The protective effect of hispidin against hydrogen peroxide-induced apoptosis in H9c2 cardiomyoblast cells through Akt/GSK-3β and ERK1/2 signaling pathway. *Exp. Cell Res.* 327(2): 264–275.

50. Penna, C., Mancardi, D., Tullio, F., and Pagliaro, P. 2008. Postconditioning and intermittent bradykinin induced cardioprotection require cyclooxygenase activation and prostacyclin release during reperfusion. *Basic Res. Cardiol.* 103: 368–377.

51. Zorov, D.B., Filburn, C.R., Klotz, L.O., Zweier, J.L., and Sollott, S.J. 2000. Reactive oxygen species (ROS)-induced ROS release: A new phenomenon accompanying induction of the mitochondrial permeability transition in cardiac myocytes. *J. Exp. Med.* 192: 1001–1014.

52. Lee, I.K., Seok, S.J., Kim, W.K., and Yun, B.S. 2006. Hispidin derivatives from the mushroom *Inonotus xeranticus* and their antioxidant activity. *J. Nat. Prod.* 69: 299–301.

53. Jang, J.S., Lee, J.S., Lee, J.H., Kwon, D.S., Lee, K.E., Lee, S.Y., and Hong, E.K. 2010. Hispidin produced from *Phellinus linteus* protects pancreatic beta-cells from damage by hydrogen peroxide. *Arch. Pharm. Res.* 33: 853–861.

54. Martin, A.R., Villegas, I., La Casa, C., and de la Lastra, C.A. 2004. Resveratrol, a polyphenol found in grapes, suppresses oxidative damage and stimulates apoptosis during early colonic inflammation in rats. *Biochem. Pharmacol.* 67: 1399–1410.

55. McPhail, D.B., Hartley, R.C., Gardner, P.T., and Duthie, G.G. 2003. Kinetic and stoichiometric assessment of the antioxidant activity of flavonoids by electron spin resonance spectroscopy. *J. Agric. Food Chem.* 51: 1684–1690.

56. Park, I.H., Chung, S.K., Lee, K.B., Yoo, Y.C., Kim, S.K., Kim, G.S., and Song, K.S. 2004. An antioxidant hispidin from the mycelial cultures of *Phellinus linteus. Arch. Pharm. Res.* 27: 615–618.

57. Chelikani, P., Fita, I., Loewen, P.C. 2004. Diversity of structures and properties among catalases. *Cell. Mol. Life Sci.* 61: 192–208.

58. Forstermann, U. 2008. Oxidative stress in vascular disease: Causes, defense mechanisms and potential therapies. *Nat. Clin. Pract. Cardiovasc. Med.* 5: 338–349.

59. Kryston, T.B., Georgiev, A.B., Pissis, P., and Georgakilas, A.G. 2011. Role of oxidative stress and DNA damage in human carcinogenesis. *Mutat. Res.* 711: 193–201.

60. Sun, X., Sun, G.B., Wang, M., Xiao, J., and Sun, X.B. 2011. Protective effects of cynaroside against H(2)O(2)-induced apoptosis in H9c2 cardiomyoblasts. *J. Cell. Biochem.* 112: 2019–2029.

61. Hausenloy, D.J. and Yellon, D.M. 2004. New directions for protecting the heart against ischaemia–reperfusion injury: Targeting the Reperfusion Injury Salvage Kinase (RISK)-pathway. *Cardiovasc. Res.* 61: 448–460.

62. Armstrong, S.C. 2004. Protein kinase activation and myocardial ischemia/reperfusion injury. *Cardiovasc. Res.* 61: 427–436.

63. Jeong, J.J., Ha, Y.M., Jin, Y.C., Lee, E.J., Kim, J.S., Kim, H.J., Seo, H.G., Lee, J.H, Kang, S.S., Kim, Y.S., and Chang, K.C. 2009. Rutin from *Lonicera japonica* inhibits myocardial ischemia/reperfusion-induced apoptosis in vivo and protects H9c2 cells against hydrogen peroxide-mediated injury via ERK1/2 and PI3K/Akt signals in vitro. *Food Chem. Toxicol.* 47: 1569–1576.

64. Yamaguchi, H. and Wang, H.G. 2001. The protein kinase PKB/Akt regulates cell survival and apoptosis by inhibiting Bax conformational change. *Oncogene* 20: 7779–7786.

65. Weston, C.R., Balmanno, K., Chalmers, C., Hadfield, K., Molton, S.A., Ley, R., Wagner, E.F., and Cook, S.J. 2003. Activation of ERK1/2 by deltaRaf-1:ER* represses Bim expression independently of the JNK or PI3K pathways. *Oncogene* 22: 1281–1293.

66. Terada, K., Kaziro, Y., and Satoh, T. 2000. Analysis of Ras-dependent signals that prevent caspase-3 activation and apoptosis induced by cytokine deprivation in hematopoietic cells. *Biochem. Biophys. Res. Commun.* 267: 449–455.

67. Cardone, M.H., Roy, N., Stennicke, H.R., Salvesen, G.S., Franke, T.F., Stanbridge, E., Frisch, S., and Reed, J.C. 1998. Regulation of cell death protease caspase-9 by phosphorylation. *Science* 282: 1318–1321.

68. Hahn, H.S., Yussman, M.G., Toyokawa, T., Marreez, Y., Barrett, T.J., Hilty, K.C., Osinska, H., Robbins, J., and Dorn, G.W. 2nd. 2002. Ischemic protection and myofibrillar cardiomyopathy: Dose-dependent effects of in vivo deltaPKC inhibition. *Circ. Res.* 91: 741–748.

69. Peterman, E.E., Taormina, P. 2nd., Harvey, M., and Young, L.H. 2004. Go 6983 exerts cardioprotective effects in myocardial ischemia/reperfusion. *J. Cardiovasc. Pharmacol.* 43: 645–656.

70. Hug, H. and Sarre, T.F. 1993. Protein kinase C isoenzymes: Divergence in signal transduction? *Biochem. J.* 291: 329–343.

71. Numaguchi, K., Shimokawa, H., Nakaike, R., Egashira, K., and Takeshita, A. 1996. PKC inhibitors prevent endothelial dysfunction after myocardial ischemia–reperfusion in rats. *Am. J. Physiol.* 270: H1634–H1639.

8 Herbal Formulations in Folk Medicine

Intensive studies on the health benefits of herbal medicine have offered new treatment strategy for various ailments. Recent studies show herbal medicines can be effective in managing allergy as well [1–4]. However, individual active phytochemicals usually do not produce considerable pharmacological activities in organisms, yet the potency of herbal medicines unravels slowly. Table 8.1 shows the impact of herbal medicines and their activities in humans. Specific herbal formulations not only have effects on the targeted organ or tissue but may also exhibit effects on the immune system, which is believed to play a crucial rule in treatment of human diseases [5–9].

MOBILIZATION OF THE IMMUNE SYSTEM WITH HERBAL MEDICINE

The immune system can be mobilized to fight against cancer as well. A recent study on *Pleurotus eryngii* Quel showed that it can be used to treat various disorders and diseases [7]. The fruiting bodies of *P. eryngii* were used for immunostimulation, wound healing, lumbago, and cancer [7]. *P. eryngii* fruiting bodies powder has a potential application as an antitumor agent with immunomodulatory activity through the inhibition of the lysosomal activity of cancerous cells. It can stimulate macrophage-mediated immune responses. There are not many herbal phytochemicals that can work as immune depressants or stimulants. Table 8.2 highlights the impact of herbal medicines that can affect immune responses.

COMPOSITION OF CHINESE MEDICINE

A basic theory of herbal composition with traditional Chinese medicines is often used to add on or subtract from the specific formulation some ingredients to enhance the pharmacological effects on biochemical functions in humans. This principle plays an important role in developing personalized medicine and individual compound compatibility of traditional Chinese medicines, yet the theory underlying the change in herbal formulas remains sketchy. The composition of herbal medicine depends on the experience of the Chinese medicine practitioners who prescribe the

TABLE 8.1
Impacts of Herbal Medicines on Human Health and Their Major Activities

Herbal Compound	Specialty	References
1. Pharmacological activity		
Rikkunshito	1. Prokinetic action on gastric emptying. 2. Therapeutic effects against gastroesophageal reflux disease and duodenogastroesophageal reflux associated with cytotoxicity of the bile salts in the gut lumen.	[15]
Methanol fraction of *Amomum xanthioides*	Inhibits collagen accumulation and activation of hepatocyte stellate cells in the liver tissue.	[16]
Zizyphus jujuba	Reduces bilirubin concentration.	[17]
Bacopasides from *Bacopa monnieri*	1. Inhibits opioid-withdrawal-induced hyperalgesia and acquisition and expression of morphine tolerance. 2. Strong protective effect against toxic effects of opiates on major organs such as brain, kidneys, and heart.	[18]
Naringenin	Prevents the accumulation of plasma lipids and lipoproteins; decreases the levels of plasma high-density lipoprotein, lipoprotein lipase, and lecithin cholesterol acyltransferase.	[19]
2. Others		
Olive leaf extract	1. Normalizes glucose homeostasis in individuals with diabetes. 2. Reduces starch digestion and absorption.	[20]
Ellagitannins	Inhibit the secretion of MMP-9 induced by hemozoin or TNF and MMP-9 promoter activity and NF-κB-driven transcription.	[21]
Ophiopogonin D	Inhibits mRNA levels of antioxidant, inflammatory, and apoptotic genes in HUVECs; H_2O_2-induced lipid peroxidation and protein carbonylation were reduced by OP-D pretreatment mitochondria.	[22]
Bupleurum falcatum	Antidepressant-like activity to alternative therapy for depression via the serotonergic and noradrenergic systems.	[23]
JD-30 from Danggui-Shaoyao-San	Prevents the aggregation of A beta (25–35), but disrupted aggregated A beta (25–35) fibrils.	[24]

(Continued)

TABLE 8.1 (*Continued*)
Impacts of Herbal Medicines on Human Health and
Their Major Activities

Herbal Compound	Specialty	References
Ginseng	Attenuates oxidative stress and reduces the NF-kappa B (p65) levels.	[25]
Herba Epimedii	1. Improves bone health and cardiovascular function. 2. Regulates hormone level and modulates immunological function. 3. Inhibits tumor growth.	[26]
Soshiho-tang	1. Treatment for hepatitis, liver cirrhosis, and chronic and acute liver diseases. 2. Antithrombotic and antiplatelet activities via inhibition of $FeCl_3$-induced thrombus formation through antiplatelet activity. 3. Inhibition of platelet aggregation and serotonin and TXB2 productions.	[27]
Heshouwuyin	Invigorates the kidney via the control of testosterone secretion and sperm function.	[28]
Tiáo-Gēng-Tāng (TG decoction)	1. Balances female hormones. 2. Regulates expression of estrogen receptors. 3. Prevents aging-related tissue damage.	[29]
Scutellarin	1. Antioxidative, anti-inflammatory, vasodilator, and cardiovascular and cerebrovascular ischemia protective effects. 2. Promotes angiogenesis.	[30]
Curcumin	Inflammatory bowel disease.	[31]
Sanshuibaihu decoction	Inhibits effect on collagen-induced arthritis associated with NF-kappa B and p38 MAPK alpha.	[32]
Butanol-purified extract of Food Allergy Herbal Formula-2	Suppresses Th2 cytokine, IgE, and histamine levels in vivo and shows direct inhibition of Th2, IgE-producing B cells, and mast cell activation in vitro.	[33]
Emodin	Inhibits OVA-induced increases in eosinophil count and interleukin (IL)-4, IL-5, and IL-13 levels; reduces serum levels of OVA-specific IgE, IgG, and IgG1.	[34]
Bacopasides from *Bacopa monnieri*	Anti-inflammatory effect via COX-2 inhibitory mechanism.	[18]

TABLE 8.2
Health Benefits of Herbal Medicine on Immunity

Herbal Compound	Specialty	References
Sanshuibaihu decoction	Treatment of rheumatoid arthritis	[32]
Triptolide	Anti-inflammatory and immunosuppressive effects in vitro and in vivo	[35]

specific composition of herbs. An integrated platform of system pharmacology revealed how the pharmacokinetics of herbal medicines and target interactions can be assessed [10]. The studies predicted the major constituents, the molecular targets, and their interaction pathways associated with specific herbal medicine. For examples, Xiao Chaihu decoction and Da Chaihu decoction were used to demonstrate the approach used to unravel their interactions [10]. This study provides a novel methodology to study the mechanism of the addition and subtraction theory in designing a new herbal formula, with desirable therapeutic effects. In addition, the activity-integrated method for both quality assessment and screening of the potential antioxidants can be assured for determination of the pharmacological activity and contents of the constituents in the herbal samples [11]. Another recent study showed the method for pharmacokinetic evaluation of five protoberberine-type alkaloids in rats after single and multiple oral administrations of Jiao-Tai-Wan [12]. The pharmacokinetic activities of the five alkaloids had significant differences in the single dose or multiple doses between normal and insomniac rats. Multiple dosing reflects enhanced absorption efficiency, which is believed to attribute to the significant therapeutic effects in insomniac rats.

APPLICATION OF CHINESE MEDICINE IN PEDIATRIC POPULATION

The use of complementary alternative medicine has become increasingly popular in pediatric populations for the management of pain and other chronic diseases [13]. The advantages of an integrated approach have aroused intensive research into integrated medicine. The exploitation of a more effective therapy strategy prompts the use of herbal medicine and Western drugs for the treatment of diseases especially chronic ailments and cancers. The significance of an integrated approach to manage chronic diseases has become recognized.

Multitarget therapeutics is a promising paradigm for drug discovery, which is believed to lead to the production of more efficacious drugs with less side effects and toxicity than monotherapy [14]. Herbal medicines contain different types of active phytochemicals that may have multitargets in the systems. Specific drug-target bioinformatics was constructed to capture the key disease-relevant biology [14]. This strategy integrating different types of technologies including a comprehensive database, phytochemicals, and pharmacology for herbal medicines and systems biology is expected to help create new opportunities for drug discovery from herbal medicines and combination therapy.

Many studies [15–18] confirm that a majority of patients undergoing cancer therapy use customized forms of complementary therapies together with dietary control, yet cancer patients often overlook the effects of food supplements, which sometimes contain herbal ingredients. The failure of Chinese medicine practitioners to communicate effectively with patients on this use may result in the treatment strategy. Unfortunately, this could be harmful or ineffective therapies when integrative interventions can be established. Perhaps the optimal approach is to discuss the treatment strategy with the patient in order to enhance the therapeutic effects of herbal medicine. With the use of medical or food supplements with herbal medicine, integrative medicine should provide basic information to physicians who offer complementary and alternative medicine for cancer therapy. Common supplements include curcumin, glutamine, vitamin D, Maitake mushrooms, fish oil, green tea, milk thistle, astragalus, melatonin, and probiotics. Basic information with an up-to-date base of knowledge, such as evidence on effectiveness and clinical trials, adverse effects, and interactions with medications, should be made available to patients. These enable physicians and patients to be aware of the supplements and potential benefits and risks of medical and food supplements that contain herbal ingredients.

REFERENCES

1. Xue, C.C., Hügel, H.M., Li, C.G., and Story, D.F. 2004. Efficacy, chemistry and pharmacology of Chinese herbal medicine for allergic rhinitis. *Curr. Med. Chem.* 11(11): 1403–1421.
2. De Smet, P.A. 2005. Herbal medicine in Europe—Relaxing regulatory standards. *N. Engl. J. Med.* 352(12): 1176–1178.
3. Shao, Z.J., Zheng, X.W., Feng, T., Huang, J., Chen, J., Wu, Y.Y., Zhou, L.M., Tu, W.W., and Li, H. 2012. Andrographolide exerted its antimicrobial effects by upregulation of human β-defensin-2 induced through p38 MAPK and NF-κB pathway in human lung epithelial cells. *Can. J. Physiol. Pharmacol.* 90(5): 647–653.
4. Kamyab, A.A. and Eshraghian, A. 2013. Anti-inflammatory, gastrointestinal and hepatoprotective effects of *Ocimum sanctum* Linn: An ancient remedy with new application. *Inflamm. Allergy Drug Targets* 12(6): 378–384.

5. Kiyohara, H., Matsumoto, T., and Yamada, H. 2002. Intestinal immune system modulating polysaccharides in a Japanese herbal (Kampo) medicine, Juzen-Taiho-To. *Phytomedicine* 9(7): 614–624.

6. Wang, H., Chan, Y.L., Li, T.L., and Wu, C.J. 2012. Improving cachectic symptoms and immune strength of tumour-bearing mice in chemotherapy by a combination of *Scutellaria baicalensis* and Qing-Shu-Yi-Qi-Tang. *Eur. J. Cancer* 48(7): 1074–1084.

7. Mariga A.M., Pei F., Yang, W.J., Zhao, L.Y., Shao, Y.N., Mugambi, D.K., and Hu, Q.H. 2014. Immunopotentiation of *Pleurotus eryngii* (DC. ex Fr.) Quel. *J. Ethnopharmacol.* 153(3): 604–614.

8. Taguchi, A., Kawana, K., Yokoyama, T., Adachi, K., Yamashita, A., Tomio, K., Kojima, S., Oda, K., Fujii, T., and Kozuma, S. 2012. Adjuvant effect of Japanese herbal medicines on the mucosal type 1 immune responses to human papillomavirus (HPV) E7 in mice immunized orally with Lactobacillus-based therapeutic HPV vaccine in a synergistic manner. *Vaccine* 30(36): 5368–5372.

9. Ren, Z., He, C., Fan, Y., Guo, L., Si, H., Wang, Y., Shi, Z., and Zhang, H. 2014. Immuno-enhancement effects of ethanol extract from *Cyrtomium macrophyllum* (Makino) Tagawa on cyclophosphamide-induced immunosuppression in BALB/c mice. *J. Ethnopharmacol.* 155(1): 769–775.

10. Li, B., Tao, W., Zheng, C., Shar, P.A., Huang, C., Fu, Y., and Wang, Y. 2014. Systems pharmacology-based approach for dissecting the addition and subtraction theory of traditional Chinese medicine: An example using Xiao-Chaihu-Decoction and Da-Chaihu-Decoction. *Comput. Biol. Med.* 53: 19–29.

11. Chang, Y.X., Liu, J., Bai, Y., Li, J., Liu, E.W., He, J., Jiao, X.C., Wang, Z.Z, Gao, X.M., Zhang, B.L., and Xiao, W. 2014. The activity-integrated method for quality assessment of reduning injection by on-line DPPH-CE-DAD. *PLoS One* 9(9): e106254. doi: 10.1371/journal.pone.0106254.

12. He, W., Liu, G., Cai, H., Sun, X., Hou, W., Zhang, P., Xie, Z., and Liao, Q. 2014. Integrated pharmacokinetics of five protoberberine-type alkaloids in normal and insomnic rats after single and multiple oral administration of Jiao-Tai-Wan. *J. Ethnopharmacol.* 154(3): 635–644.

13. Dalla Libera, D., Colombo, B., Pavan, G., and Comi, G. 2014. Complementary and alternative medicine (CAM) use in an Italian cohort of pediatric headache patients: The tip of the iceberg. *Neurol. Sci.* 35(1): 145–148.

14. Li, P., Chen, J., Wang, J., Zhou, W., Wang, X., Li, B., Tao, W., Wang, W., Wang, Y., and Yang, L. 2014. Systems pharmacology strategies for drug discovery and combination with applications to cardiovascular diseases. *J. Ethnopharmacol.* 151(1): 93–107.

15. Araki, Y., Mukaisho, K.I., Fujiyama, Y., Hattori, T., and Sugihara, H. 2012. The herbal medicine rikkunshito exhibits strong and differential adsorption properties for bile salts. *Exp. Ther. Med.* 3: 645–649.

16. Wang, J.H., Shin, J.W., Choi, M.K., Kim, H.G., and Son, C.G. 2011. An herbal fruit, *Amomum xanthioides*, ameliorates thioacetamide-induced hepatic fibrosis in rat via antioxidative system. *J. Ethnopharmacol.* 135: 344–350.

17. Ebrahimi, S., Ashkani-Esfahani, S., and Poormahmudi, A. 2011. Investigating the efficacy of *Zizyphus jujuba* on neonatal jaundice. *Iran. J. Pediatr.* 21: 320–324.
18. Rauf, K., Subhan, F., Al-Othman, A.M., Khan, I., Zarrelli, A., and Shah, M.R. 2013. Preclinical profile of bacopasides from *Bacopa monnieri* (BM) as an emerging class of therapeutics for management of chronic pains. *Curr. Med. Chem.* 20: 1028–1037.
19. Jayachitra, J., and Nalini, N. 2012. Effect of naringenin (citrus flavanone) on lipid profile in ethanol-induced toxicity in rats. *J. Food Biochem.* 36: 502–511.
20. Wainstein, J., Ganz, T., Boaz, M., Bar Dayan, Y., Dolev, E., Kerem, Z., and Madar, Z. 2012. Olive leaf extract as a hypoglycemic agent in both human diabetic subjects and in rats. *J. Med. Food* 15: 605–610.
21. Dell'Agli, M., Galli, G.V., Bulgari, M., Basilico, N., Romeo, S., Bhattacharya, D., Taramelli, D., and Bosisio, E. 2010. Ellagitannins of the fruit rind of pomegranate (*Punica granatum*) antagonize in vitro the host inflammatory response mechanisms involved in the onset of malaria. *Malaria J.* 9: 208.
22. Qian, J.C., Jiang, F.R., Wang, B., Yu, Y., Zhang, X., Yin, Z.M., and Liu, C. 2010. Ophiopogonin D prevents H2O2-induced injury in primary human umbilical vein endothelial cells. *J. Ethnopharmacol.* 128: 438–445.
23. Kwon, S., Lee, B., Kim, M., Lee, H., Park, H.J., and Hahm, D.H. 2010. Antidepressant-like effect of the methanolic extract from *Bupleurum falcatum* in the tail suspension test. *Prog. Neuropsychopharmacol. Biol. Psychiatry* 34: 265–270.
24. Hu, Z.Y., Liu, G., Yuan, H., Yang, S., Zhou, W.X., Zhang, Y.X., and Qiao, S.Y. 2010. Danggui-Shaoyao-San and its active fraction JD-30 improve a beta-induced spatial recognition deficits in mice. *J. Ethnopharmacol.* 128: 365–372.
25. Sen, S., Chen, S.L., Feng, B.A., Wu, Y.X., Lui, E., and Chakrabarti, S. 2012. Preventive effects of North American ginseng (*Panax quinquefolium*) on diabetic nephropathy. *Phytomedicine* 19: 494–505.
26. Zhai, Y.K., Guo, X., Pan, Y.L., Niu, Y.B., Li, C.R., Wu, X.L., and Mei, Q.B. 2013. A systematic review of the efficacy and pharmacological profile of Herba Epimedii in osteoporosis therapy. *Pharmazie* 68: 713–722.
27. Lee, J.J., Kim, T., Cho, W.K., and Ma, J.Y. 2013. Antithrombotic and antiplatelet activities of Soshiho-tang extract. *BMC Complement. Altern. Med.* 13: 137.
28. Niu, S., Chen, J., Duan, F., Song, Q., Qin, M., Wang, Z., and Liu, J. 2014. Possible mechanism underlying the effect of Heshouwuyin, a tonifying kidney herb, on sperm quality in aging rats. *BMC Complement. Altern. Med.* 14: 250.
29. Xu, L.W., Kluwe, L., Zhang, T.T., Li, S.N., Mou, Y.Y., Sang, Z., Ma, J., Lu, X., and Sun, Z.J. 2011. Chinese herb mix Tiao-Geng-Tang possesses antiaging and antioxidative effects and upregulates expression of estrogen receptors alpha and beta in ovariectomized rats. *BMC Complement. Altern. Med.* 11: 137.

30. Gao, Z., Huang, D., Li, H., Zhang, L., Lv, Y., Cui, H., and Zheng, J. 2010. Scutellarin promotes in vitro angiogenesis in human umbilical vein endothelial cells. *Biochem. Biophys. Res. Commun.* 400: 151–156.

31. Ali, T., Shakir, F., and Morton, J. 2012. Curcumin and inflammatory bowel disease: Biological mechanisms and clinical implication. *Digestion* 85: 249–255.

32. Yang, M., Xiao, C., Wu, Q., Niu, M., Yao, Q., Li, K., Chen, Y., Shi, C., Chen, D., Feng, G., and Xia, C. 2010. Anti-inflammatory effect of Sanshuibaihu decoction may be associated with nuclear factor-kappa B and p38 MAPK alpha in collagen-induced arthritis in rat. *J. Ethnopharmacol.* 127: 264–273.

33. Srivastava, K., Yang, N., Chen, Y., Lopez-Exposito, I., Song, Y., Goldfarb, J., Zhan, J., Sampson, H., and Li, X.M. 2011. Efficacy, safety and immunological actions of butanol-extracted Food Allergy Herbal Formula-2 on peanut anaphylaxis. *Clin. Exp. Allergy* 41: 582–591.

34. Chu, X., Wei, M.M., Yang, X.F., Cao, Q.J., Xie, X.X., Guan, M.F., Wang, D.C., and Deng, X.M. 2012. Effects of an anthraquinone derivative from Rheum officinale Baill, emodin, on airway responses in a murine model of asthma. *Food Chem. Toxicol.* 50: 2368–2375.

35. Han, R., Rostami-Yazdi, M., Gerdes, S., and Mrowietz, U. 2012. Triptolide in the treatment of psoriasis and other immune-mediated inflammatory diseases. *Br. J. Clin. Pharmacol.* 74: 424–436.

9 Exploration of Herbal Medicine

Use of plant or herbal extracts or herbal formulations to treat human diseases is very common in various folk medicines in Asian countries, especially in China medicinal herbs are known for their multifaceted pharmacological activities and thus can form an effective treatment method. Quite a number of herbal extracts or decoration phytochemicals and herbal formulations have been evaluated systematically and clinically for their therapeutic benefits in experimental animal models and in humans, but clinical studies using the same approach are sparse. Herbal extracts with antioxidants, antidiabetic, and antihyperlipidemic properties have been intensively investigated studies in the last decades.

COMPLEMENTARY AND ALTERNATIVE MEDICINE

The use of complementary and alternative medicine is widely spread not only in Asian countries, but also in the Western world [1]. Despite the increasing evidence on the harmful effects induced by herbal medicines, patients appreciate the health benefits more than the side effects of herbal medicines. As a result, there is an increasing request for experimental data to assess both the efficacy and safety of these herbal medicines. This is to support the use of such medications as adjuvant treatments to therapeutic drugs; thus, the current evidence on efficacy and safety of some natural products and herbal formulations that are believed to be effective in treatment of diseases should be provided. Further perspectives for the chemical use of herbal products and strategies for improving knowledge about herbal medicines can be refined.

PHARMACOLOGY MODEL FOR HERBAL MEDICINE INJECTION

Compared with the traditional administration form of herbal medicines, injection administration seems more effective in terms of both biological availability and therapeutic benefits [2]. However, an injection administration of herbal medicines may induce undesirable side interactions and undermining the therapeutic effects of herbal constituents. A new approach based on a novel pharmacology model integrating absorption distribution, metabolism, and excretion (ADME), target sites, and biochemical pathways is developed to explore the effectiveness of an injection administration to the influenza. The ADME activities helped identify 31 phytochemicals and 4 of their metabolites from the Reduning injection [2]. The methodology revealed a new way of confronting influenza disease through stimulating the immune-modulatory agents for immune response activation and regulating the inflammatory activity associated with anti-inflammation. The novel pharmacology advances the understanding of the mechanism of actions of herbal constituents in the herbal formulation and the treatment method. It can provide insights into discovering more effective drugs against complex diseases.

Herbal medicines have been used for hundreds of years for the treatment of various ailments in China and other countries in the world. Herbal medicines, especially in formulations, have multiarrays of pharmacologic activities such as improvement of blood circulation, modulation of coronary vasodilation and protection against myocardial ischemia, suppression of platelet adhesion and aggregation, and inhibition of cancer proliferation. Specific herbal formulations alone or in combination with other Western medicines show different health benefits to humans. Individual herbs can be used in combination with other herbal ingredients to enhance the pharmacological effects. Anticancer herbal formulations can act on the immune responses or multimolecular targets for curative approach to cancer therapy. The combination therapy with Western medicine could provide a more effective cancer therapy with less drug toxicity or reduction of side effects of cancer drugs.

APPLICATION OF CHINESE HERBAL MEDICINES

The emerging development of extractive reference substance (ERS) is a practical methodology that meets the requirements for quality control for Chinese herbal medicines, and its clinical use of multiple herbs with medicinal effects. The previous use of a selected few chemical reference substances (CRSs) cannot adequately evaluate the therapeutic effects of Chinese herbal medicines. A database of chemical spectrum of an ERS provides the overall characteristics of herbal medicines to enable the assessment of targeted Chinese herbal medicine [3–5].

Two common kinds of reference substances applied to Pharmacopoeia of People's Republic of China and CRS. Practical experience has demonstrated the roles of herbal reference substances and CRSs in quality control of Chinese herbal medicines. However, it is unsatisfactory for chemical identification due to variation in the method sensitivity and less specific for some herbal components.

Any kind of reference substance must meet the four basic requirements in chemical testing: authenticity, specificity, consistency, and stability.

Therefore, meeting all the requirements is very important. In order to fulfill the requirements, the ERS for identification and for full functionality in terms of qualitative and quantitative needs to take into account all four criteria in assessing Chinese herbal medicines. Although there is a fundamental difference in outlook between the Western approach of chemical pharmaceuticals and the traditional orientation of Chinese herbal medicines toward evaluating the therapeutic benefits in the human body's function, updated strategies of good quality control should be considered in assessing the complexity of Chinese herbal medicines. No work into the drug safety and efficacy can compromise the drug actions exerted by multicomponents in an herbal formulation. Research and application of ERS for a given herbal species are becoming a new trend for reference substances for the quality control of herbal medicine. The requirements for the ERS of Chinese herbal medicines should take into account of the authenticity, specificity, consistency, and stability in a systemic approach. This will enable a more effective quality control of Chinese herbal medicines.

COMPLEMENTARY APPROACH WITH HERBAL FORMULATION

Herbal formula in some variations seems commonly used in formulating Chinese medicines for the treatment of ailments. Most often, an herbal formula shows effects on patients. Furthermore, many of the herbs included in the formulas may show synergistic effects. Hence, future research should focus on testing the herbal formula with different herbal compositions. The positive effects of an herbal formula provide a good starting point in an animal study. The main problem with assessment of herbal medicines in that most studies did not use internationally recognized and validated scales. As a result, this makes comparative studies difficult and unreliable [6]. Validated scales can be modified to accommodate the assessment of Chinese medicinal formula.

Another issue in assessing efficacy and safety of traditional medicines, including complementary and alternative modalities, is that these are usually traditional, and treatment interventions are highly individualized and lacking standardization. Herbal therapies are highly tailored to the individual patients. It is difficult to assess which particular herb or combination of herbs shows therapeutic effects for a specific condition. Chinese medicine practitioners count on the synergistic effects of the combination of herbs, and unlike Western drugs that are designed for one molecular target with specific pharmacological uses, herbal medicines have a more broad spectra of actions. In addition, the herbs are used together with various forms of exercises and herbal modalities such as acupuncture.

Individualized and highly customized treatments are possible with herbal medicines. Furthermore, the skills and experience of practitioners may play a crucial role for effective treatment patients, although it is believed that there are methods to model and formulate an herbal composition for patients.

FINGERPRINT FOR HERBAL MEDICINES

Traditional Chinese medicine (TCM)-based herbal medicines have been widely accepted worldwide. Some of the TCMs are being pursued by pharmaceutical companies as resources for novel drug discovery [7–9]. Herbal remedies have been used as a complementary and alternative modality by patients, including cancer patients. There are quite a number of anticancer agents that are identified in TCM. Some of these anticancer chemicals have been used in food or medical supplements as chemo-preventive agents. Metabolic profiling provides a useful tool to assess safety and efficacy of herbal medicine products [10]. Primary metabolites in herbal medicines are often ignored to set the corresponding quality criteria in China Pharmacopoeia and national standard conventional methods, including chromatography and electrophoresis are limited to the separation and the detection method for trivial amount of metabolites. To delineate various types of metabolites in herbal medicines, different types of fingerprints are needed for a generic quality evaluation, which can be difficult due to variation in industry processing and consistency in chemical analysis. Proton nuclear magnetic resonance (1HNMR) spectroscopy provides a feasible approach to detect all proton-leaving compounds [10,11]. 1HNMR method achieved the identification and quantification of active compounds and their metabolites. 1HNMR facilitates high-throughput analysis for metabolic studies and quality control of various herbal medicines. Since the complexity of chemical composition of herbal formulation, herbal medicines cannot be completely represented by a limited number of certain bioactive phytochemicals. Principal component analysis (PCA) [12] and independent component analysis (ICA) [13] are tools used to reduce the dimension of multivariate ICA has been found to be a successful alternative to PCA in eliminating the overlapping information between the components [14]. However, the methodology is limited by the number of components in the extract.

TRADE-OFFS OF HERBAL RESEARCH

In the field of complementary medicine research, herbal medicine or phytotherapy has widely been accepted in treatment of diseases. They have become the origin of new synthetic medicines [15]. They constitute the basis of patentable extracts, which can be commercially justified. As medicinal therapies, they offer double-blind controlled clinical trial designs. Subsequently, there are large numbers of research papers on herbal remedies in the literature. The literature on Ginkgo has already produced more than 80 monographs on the herbal products in the form of drug dossiers. However, there are some concerns for the future, due to patchy record of the product with little clinical application. The high costs of conducting clinical trials to modern ethical standards and regulatory requirements disable the development prospects. A cultural difference discourages the clinical research. Sales in the largest herbal market, the United States, are based solely on the status for herbs a dietary supplements, where clinical evidence is useful but not required. There is no product protection for such investment. Better hopes lie with the sustainable economic development in countries worldwide. Renewed interest in the potential of folk medications has increased research activities in herbal medicines and plants. New standards can be employed to improve the quality of investigation and new products. New ways should be developed to service the demand of novel drug development.

Highly distinctive patterns of association between phytochemical class and profile of herbal medicine suggested that a strong phytochemical basis underlies the possible mechanism of actions of herbal medicine [16]. Relationships between profile of herbal medicines and activities that are associated with therapeutically important molecular targets can be explored. The ethnopharmacological information could play an important role in pharmaceutical prospecting from herbal medicines and characterizing links between herbal medicines, especially TCM and Western medicine.

COMBINATION THERAPY WITH WESTERN MEDICINE

The use of TCM in combination with Western medicine has become increasingly important for the treatment of diseases. It remains a widely used alternative medicine in most Asian countries. The multiarrays of pharmacological activities of TCM showcase the tremendous advantages of personalized medicine. However, personalized medicine with herbal medicine has shortcomings in the era of Western medicine that stem from its reliance on experimentation and statistical analysis. Nevertheless, the pharmacological activities of herbal medicine can be proven with an appropriate system biology and clinical study. System biology offers the potential to personalize medicine and is expected to have a major impact on the therapeutic approaches to treat numerous ailments including human cancers, yet the potential limitations of integrative medicine are needed to overcome.

REMEDY FOR DEMENTIA WITH HERBAL MEDICINE

The existing remedies for dementia in the ageing population offer limited benefits with Western medicine that causes undesirable side effects; thus, complementary and alternative medicines are being explored. China has a long history of folk medicine usage. Some of these herbal preparations are effective in enhancing blood circulation, revitalizing energy production, and resisting ageing, which are believed to be associated with alleviating symptoms of dementia [17]. These herbal formulations offer new hope for the treatment of dementia. Numerous active and potent phytochemicals can be identified from these herbal extracts. TCM offers wide applications in clinical treatment. However, more systematic study and statistics are warranted to provide supportive clinical data on herbal medicines. The use of folk medicine formulations for the treatment of diseases has aroused a lot of interest and research worldwide, yet a theoretical basis with herbal formulations for clinical treatment of human diseases is needed. For example, TCM was reported to improve osteoblast proliferation and differentiation by regulating the genes and OPG/RANK/RANKL signaling pathways of osteoblasts [18]. However, the detail mechanism of actions of active constituents of TCM in promoting osteoblast proliferation and differentiation remains sketchy.

MODERN CONCEPTS OF CHINESE MEDICINE

Chinese medicine emerges out of interaction between human biology, economic progress, and culture. Chinese medicine practices depend on the interaction among political, social, and economic aspects in a country. Oftentimes than not, the outcome of the interaction has little to do with science and medicine itself. TCM has been improved through continuous scientific study and clinical use in at least one-quarter the world's population. The people of China have used it to treat and prevent diseases including cancer, as well as to improve health.

In Western medicine, the mechanistic study and statistical analysis prevail, whereas in the traditional medicine of China, the qualitative assessment of individual patients such as dynamic ecosystems and the balance of "yin–yang" is important. TCM emphasizes the relationship between the cause and the consequence [19]. Based on this approach, TCM has been used for the treatment of tumors and ulcers more than 3000 years ago. Over the course of these millennia, different types of therapies with TCM have been developed. This ranges from

1. Pain reduction
2. Anti-inflammatory action
3. Reduction of tumor mass

4. Enhancement of host immune response
5. Potentiation of effects of radiation and chemotherapies
6. Prevention and amelioration of the adverse effects of therapeutic drugs
7. Reduction of oxidative stress

A previous study in 1999 already indicated that 72% of women with breast cancer used at least one form of herbal medicine but continued to use conventional treatment [19]. An understanding of herbal medicine may lead to comfort with the combination therapies. Herbal medicine appears to ameliorating the undesirable effects of chemotherapy and radiation. It improves the quality of life of cancer patients through the actions multifacets of herbal medicine toward different organs. With advent of the Chinese medical science, herbal medicine has become source of novel drugs. There are various approaches to the subject as how TCM can be used effectively to treat cancer in humans and animal studies.

TRADITIONAL CHINESE MEDICINE AND CANCER

In herbal medicine, mutual injury is a result of disorganization imbalance of "yin–yang" (Figure 9.1). To Western culture, this is not easy to understand how it works. Cancerous masses are the consequences of adverse accumulations of moisture and blood in Chinese perspective that have become detrimental. Prolonged stagnation can lead to depletion of blood and essence, which plays an important role in cellular formation, resulting

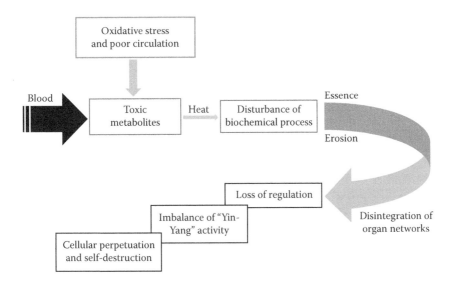

FIGURE 9.1   Pathogenesis of cancer.

cancer development; thus, treatment of the unmitigated accumulations can restore circulation and ameliorating toxic actions and removing toxic metabolites.

Both benign and malignant tumor masses can be differentiated as ulcers that appear on the surface of tissue such as muscles and skin.

Classical and modern findings regard this sort of disorders as internal injuries including benign and malignant tumors. Tumors are the consequence of prolonged process of accumulation and abnormal tissue development due to stagnation of "qi" and blood.

The relationship between generating blood and its effective circulation can improve deficiency and stagnation. The therapeutic measures with TCM are used to remove the accumulation of toxic metabolites and generate new blood. The view of Chinese medicine does not emphasize on the effects of oxidative stress regardless of their origins. Breast cancer may result from toxic accumulation and stagnation of "qi" and blood in the circulation. The malignant process is characterized by disorders of various systems. Deregulation of cellular functions is a manifestation of disturbance of basic cellular activity and biochemical processes, which may require a mixture of phytochemicals rather than a single therapeutic agent to modulate the harmful effects and the carcinogenic activity.

Cancer is condition of a complex malfunctioning of systems; thus, a single compound may not be effective enough to restore the vital cellular activities.

Cancer patterns involve oxidative stress and appropriate treatment to alternate the oxidative stress. The combination therapy with therapeutic drugs and TCMs show the combination therapy may offer a better therapy strategy.

The adverse effects of conventional treatment methods including radiation and chemotherapy come with the symptoms of deficiency in "qi," blood, and degeneration of health. Therefore, a demand for satisfactory reduction of these symptoms prompts the development of the therapy strategy with herbal medicine.

ACUPUNCTURE WITH HERBAL MEDICINE

The Nation Institutes of Health Consensus Development Conference in 1997 reported that there is clear evidence that acupuncture and Chinese medicine is an effective modality [19]. Acupuncture harmonizes "yin–yang" and the system of the human body. Normalization of the physiological processes plays a critical role in the prevention of diseases, including cancer.

In modern herbal medicine research with the advent of TCM in the twentieth century, scientists began to search for ways to improve outcomes for cancer patients undergoing radiation and chemotherapy. This approach can strengthen the functions of organs and reduce the side effects of drugs.

TREATMENT STRATEGIES

The formulation of Chinese herbal medicines is usually designed and used to address various deficiencies in biochemical functions. Herbal formulas may contain and emphasize one or all of the therapeutic principles. The clinical application of these principles can be illustrated by the actions and the pharmacological properties of herbs of the formula. The anticancer effects with the added advantages can enhance conventional cancer therapy. However, an herbal formula is complex but can produce therapeutic benefits. The adverse effects of conventional cancer therapy produce harmful effects including decreased resistance, reduced appetite and weight loss, and cognitive decline and musculoskeletal stiffness. An herbal formula can treat these conditions and restore healthy conditions. Such formulas show many of the promising anticancer phytochemicals.

Modern herbal formulas are often established based on the classical formulas as a function, while additional herbs with anticancer effects are included to enhance the anticancer effects of radiation and chemotherapy. Some patients who received Chinese herbal medicine combined with conventional Western cancer therapy showed an improved survival rate [19].

The preponderance of herbal medicines used for cancer has become an integral part of conventional practice for centuries in China. In addition to these venerable herbal medicines, new species have been reported and tested through modern research. As modern scientists are used to analyze single phytochemicals and to identify active compounds, intensive research has focused on herbs with anticancer properties. A multiherbal formula is more complex than a single herb resulting in producing more therapeutic benefits than one single herb alone.

Many of the herbal formulas used in modern cancer therapy and research in China are described in the Chinese Pharmacopoeia. Such formulas share many of the promising anticancer agents, commonly formed in herbs with anticancer properties.

HERBAL MEDICINES TO ENHANCE CONVENTIONAL PROTOCOLS

To overcome the adverse effects while simultaneously potentiating the therapeutic effects of drug treatment, a method by the name of all-inclusive great tonifying decoction (Shi Quan Da Bu Tang) is often used [19]. It appears to restore blood function and increase interleukin production. The formula contains the following ingredients:

1. *Panax ginseng*
2. *Angelica sinensis*
3. *Poria cocos*
4. *Atractylodes macrocephala*
5. *Astragalus membranaceus*
6. *Ligusticum wallichii*
7. *Paeonia lactiflora*
8. Prepared *Rehmannia glutinosa*
9. *Cinnamomum cassia*
10. Prepared *Glycyrrhiza uralensis*

This formula was formed to potentiate the therapeutic activity of chemotherapy and radiotherapy.

Other diseases such as traumatic brain injury, especially for the brain functional recovery after brain injury, are difficult to cure. However, Huayu capsule that contains TCM has a wide prospect for the treatment of brain injury [20]. The study showed the injured areas of brain tissue were decreased with an increase in the nerve cells adjacent to the injured area. The average gray degree of Nissl body in the cytoplasm of nerve cells after Huayu capsule treatment was significantly increased.

In conclusion, herbal medicines including TCM can provide an excellent complementary and alternative medicine for the treatment of numerous ailments including cancers in humans, yet more extensive studies on the pharmacological activities and the mechanism of actions of herbal medicine especially the active phytochemicals are needed before integrative medicine with Western medicine can be established for personalized medicine.

REFERENCES

1. Gilardi, D., Fiorino, G., Genua, M., Allocca, M., and Danese, S. 2014. Complementary and alternative medicine in inflammatory bowel diseases: What is the future in the field of herbal medicine? *Expert Rev. Gastroenterol. Hepatol.* 8(7): 835–846.
2. Yang, H., Zhang, W., Huang, C., Zhou, W., Yao, Y., Wang, Z., Li, Y., Xiao, W., and Wang, Y. 2014. A novel systems pharmacology model for herbal medicine injection: A case using Reduning injection. *BMC Complement. Altern. Med.* 14: 430.

3. Xie, P., Ma, S., Tu, P., Wang, Z., Stoeger, E., and Bensky, D. 2013. The prospect of application of extractive reference substance of Chinese herbal medicines. *Chin. Med.* 4: 125–136.

4. Li, S., Han, Q., Qiao, C., Song, J., Lung Cheng, C., and Xu, H. 2008. Chemical markers for the quality control of herbal medicines: An overview. *Chin. Med.* 3(7): 1–16.

5. Pang, Y., Sun, L., Jin, H., and Ma, S. 2013. Discussion on application and technical requirements of substitute reference substance method for simultaneous determination of multi-components in traditional Chinese medicine. *Chin. J. Pharm. Anal.* 33(1): 169–177.

6. Molassiotis, A., Potrata, B., and Cheng, K.K. 2009. A systematic review of the effectiveness of Chinese herbal medication in symptom management and improvement of quality of life in adult cancer patients. *Complement. Ther. Med.* 17(2): 92–120.

7. Agarwal, N., Chandra, A., and Tyagi, L.K. 2011. Herbal medicine: Alternative treatment for cancer therapy. *Int. J. Pharm. Sci. Res.* 2(9): 2249–2258.

8. Ahmadi, A., Habibi, G., and Farrokhnia, M. 2010. Anticancer effects of HESA-A: An herbal marine compound. *Chin. J. Integr. Med.* 16(4): 366–367.

9. Bachmeier, B.E., Mirisola, V., Romeo, F., Generoso, L., Esposito, A., Dell'eva, R., Blengio, F., Killian, P.H., Albini, A., and Pfeffer, U. 2010. Reference profile correlation reveals estrogen-like transcriptional activity of Curcumin. *Cell Physiol. Biochem.* 26(3): 471–482.

10. Jiang, M., Jiao, Y., Wang, Y., Xu, L., Wang, M., Zhao, B., Jia, L., Pan, H., Zhu, Y., and Gao, X. 2014. Quantitative profiling of polar metabolites in herbal medicine injections for multivariate statistical evaluation based on independence principal component analysis. *PLos One* 9(8): e105412.

11. Chen, X., Lou, Z., Zhang, H., Tan, G., Liu, Z., Li, W., Zhu, Z., and Chai, Y. 2011. Identification of multiple components in Guanxinning injection using hydrophilic interaction liquid chromatography/time-of-flight mass spectrometry and reversed-phase liquid chromatography/time-of-flight mass spectrometry. *Rapid Commun. Mass Spectrom.* 25(11): 1661–1674.

12. Jolliffe, I. 2002. *Principle Component Analysis*, 2nd edn. Springer, New York.

13. Comon, P. 1994. Independent component analysis, A new concept? *Signal Processing* 36: 287–314.

14. Hyvärinen, A., Karhunen, J., and Oja, E. 2001. *Independent Component Analysis*, 1st edn. John Wiley & Sons, New York.

15. Mills, S.Y. 2007. Herbal research: The good, the bad and the worrying. *Complement. Therap. Med.* 15(1): 1–2.

16. Ehrman, T.M., Barlow, D.J., and Hylands, P.J. 2007. Phytochemical informatics of traditional Chinese medicine and therapeutic relevance. *J. Chem. Inf. Model.* 47(6): 2316–2334.

17. Wang, Y., Huang, L.Q., Tang, X.C., and Zhang, H.Y. 2010. Retrospect and prospect of active principles from Chinese herbs in the treatment of dementia. *Acta Pharm. Sin. A* 31(6): 649–664.

18. Chen, S.X. and Kang, Le. 2012. Symbol mechanism and effect of traditional Chinese medicine on promoting osteoblast proliferation and differentiation. *J. Clin. Rehabil. Tissue Eng. Res.* 16(7): 1299–1302.
19. Beinfield, H. and Korngold, E. 2003. Alternative therapies in health and medicine. *Chin. Med. Cancer Care* 9(5): 38–52.
20. Zhou, C., Zhang, J., Wang, Y., Qian, H., Gong, L., and Huang, G. 2007. Huayu capsule enhances limb-catching capability of rats with experimental open traumatic brain injury. *Neural Regen. Res.* 2(4): 221–224.

Appendix: List of Selected Phytochemicals

Name:	Andrographolide
Chemical formula:	3-[2-[decahydro-6-hydroxy-5-(hydroxymethyl)-5,8a-dimethyl-2-methylene-1-napthalenyl]ethylidene]dihydro-4-hydroxy-2(3H)-furanone, $C_{20}H_{30}O_5$
Physical properties:	Rhombic prisms or plates from ethanol or methanol
Sources:	*Andrographis paniculata*
Therapeutic effects:	Anticancer
Chemical structure:	

Name:	Berberine
Chemical formula:	Umbellatine, 5,6-dihydro-9,10-dimethoxybenzo[g]-1,3-benzodioxolo[5,6-a]quinolizinium, $C_{20}H_{18}NO_4^+$
Physical properties:	Yellow solid
Sources:	*Berberis* [e.g., *Berberis aquifolium* (Oregon grape), *Berberis vulgaris* (barberry)]
Therapeutic effects:	Antimicrobial, anti-inflammatory, lower blood glucose level (treatment for diabetes)
Chemical structure:	

Name:	Celastrol
Chemical formula:	3-Hydroxy-9β,13α-dimethyl-2-oxo-24,25,26-trinoroleana-1(10),3,5,7-tetraen-29-oic acid, $C_{29}H_{38}O_4$
Physical properties:	Crystalline solid
Sources:	Root extracts of *Tripterygium wilfordii* (Thunder god vine) and *Celastrus regelii*
Therapeutic effects:	Anti-inflammatory, immunologic, and antiallergenic effects
Chemical structure:	

Name:	Chlorogenic acid
Chemical formula:	(1S,3R,4R,5R)-3-{[(2E)-3-(3,4-dihydroxyphenyl)prop-2-enoyl]oxy}-1,4,5-trihydroxycyclohexanecarboxylic acid, $C_{16}H_{18}O_9$
Physical properties:	Hemihydrate, needles from water
Sources:	*Phyllostachys edulis*, green coffee bean extract
Therapeutic effects:	Antioxidant
Chemical structure:	

Name:	Coumarin
Chemical formula:	2H-chromen-2-one, $C_9H_6O_2$
Physical properties:	Sweet odor, colorless crystalline substance
Sources:	Tonka beans
Biologic activities:	Precursor reagent for synthetic anticoagulant pharmaceuticals, edema modifier
Chemical structure:	

Name:	Cryptochlorogenic acid
Chemical formula:	$C_{16}H_{18}O_9$ (derivative of chlorogenic acid)
Physical properties:	White crystalline
Sources:	Roasted coffee beans, radix *Salvia miltiorrhiza* (danshen)
Biologic activities:	Antioxidant
Chemical structure:	

Name:	Friedelin
Chemical formula:	(4*R*,4a*S*,6a*S*,6a*S*,6b*R*,8a*R*,12a*R*,14a*S*,14b*S*)-4,4a,6a,6b,8a,11,11,14a-octamethyl-2,4,5,6,6a,7,8,9,10,12,12a,13,14,14b-tetradecahydro-1*H*-picen-3-one, $C_{30}H_{50}O$
Physical properties:	Needles from ethyl acetate or alcohol
Sources:	*Azima tetracantha*, *Orostachys japonica*, and *Quercus stenophylla*
Therapeutic effects:	Anti-inflammatory, analgesic, and antipyretic
Chemical structure:	

Name:	Gambogic acid
Chemical formula:	(Z)-4-((1S,3aR,5S,11R,14aS)-8-hydroxy-2,2,11-trimethyl-13-(3-methylbut-2-en-1-yl)-11-(4-methylpent-3-en-1-yl)-4,7-dioxo-2,3a,4,5,7,11-hexahydro-1H-1,5-methanofuro[3,2-g]pyrano[3,2-b]xanthen-3a-yl)-2-methylbut-2-enoic acid, $C_{38}H_{44}O_8$
Physical properties:	Yellow prisms from methanol
Sources:	*Garcinia hanburyi*
Therapeutic effects:	Antitumor
Chemical structure:	

Name:	Icariin
Chemical formula:	5-hydroxy-2-(4-methoxyphenyl)-8-(3-methylbut-2-enyl)-7-[(2S,3R,4S,5S,6R)-3,4,5-trihydroxy-6-(hydroxymethyl)oxan-2-yl]oxy-3-[(2S,3R,4R,5R,6S)-3,4,5-trihydroxy-6-methyloxan-2-yl]oxychromen-4-one, $C_{33}H_{40}O_{15}$
Physical properties:	Crystalline solid
Sources:	*Epimedium*, *Berberidaceae*
Therapeutic effects:	Aphrodisiac, antioxidant
Chemical structure:	

Name:	Isoliquiritigenin
Chemical formula:	(*E*)-1-(2,4-dihydroxyphenyl)-3-(4-hydroxyphenyl)prop-2-en-1-one, $C_{15}H_{12}O_4$
Physical properties:	Yellow powder
Sources:	Licorice
Therapeutic effects:	Anticancer
Chemical structure:	

Name:	Isoliquiritin
Chemical formula:	(*E*)-1-(2,4-dihydroxyphenyl)-3-[4-[(2*S*,3*R*,4*S*,5*S*,6*R*)-3,4,5-trihydroxy-6-(hydroxymethyl)oxan-2-yl]oxyphenyl]prop-2-en-1-one, $C_{21}H_{22}O_9$
Physical properties:	White powder
Sources:	Licorice
Therapeutic effects:	Anticancer
Chemical structure:	

Name:	Leonurine
Chemical formula:	4-Hydroxy-3,5-dimethoxybenzoic acid 4-guanidinobutyl ester, $C_{14}H_{21}N_3O_5$
Physical properties:	White powder
Sources:	*Leonotis leonurus, Leonotis nepetifolia, Leonotis artemisia, Leonurus cardiaca* (motherwort), *Lamiaceae, Leonurus sibiricus*
Therapeutic effects:	Antioxidant
Chemical structure:	

Name: Ligustrazine (tetramethylpyrazine)
Chemical formula: 2,3,5,6-Tetramethylpyrazine, $C_8H_{12}N_2$
Physical properties: White crystals
Sources: Natto (soybeans fermented with *Bacillus subtilis* var. *natto*)
Therapeutic effects: Anti-inflammatory
Chemical structure:

Name: Liquiritin
Chemical formula: (2S)-7-hydroxy-2-[4-[(2S,3R,4S,5S,6R)-3,4,5-trihydroxy-6-
                                 (hydroxymethyl)oxan-2-yl]oxyphenyl]-2,3-dihydrochromen-
                                 4-one, $C_{21}H_{22}O_9$
Physical properties: Powder
Sources: Licorice
Therapeutic effects: Anticancer
Chemical structure:

Name: Neochlorogenic acid
Chemical formula: (1R,3R,4S,5R)-3-{[(2E)-3-(3,4-Dihydroxyphenyl)prop-2-
                                 enoyl]oxy}-1,4,5-trihydroxycyclohexanecarboxylic acid,
                                 $C_{16}H_{18}O_9$
Physical properties: White crystalline
Sources: Coffee, prunes, bamboo *Phyllostachys edulis*, Radix *Salvia
                                 miltiorrhiza*
Biologic activities: Antioxidant
Chemical structure:

Name:	Oleanolic acid
Chemical formula:	(4aS,6aR,6aS,6bR,8aR,10S,12aR,14bS)-10-hydroxy-2,2,6a,6b,9,9,12a-heptamethyl-1,3,4,5,6,6a,7,8,8a,10,11,12,13,14b-tetradecahydropicene-4a-carboxylic acid, $C_{30}H_{48}O_3$
Physical properties:	Fine, solvated needles from alcohol
Sources:	*Olea europaea, Oleaceae, Phytolacca americana* (American pokeweed), *Syzygium* spp.
Therapeutic effects:	Hepatoprotective, antitumor
Chemical structure:	

Name:	Oxymatrine
Chemical formula:	(7aS,13aR,13bR,13cS)dodecahydro-1H,5H,10H-dipyrido[2,1-f:3′,2′,1′-ij][1,6]naphthyridin-10-one 4-oxide, $C_{15}H_{24}N_2O_2$
Physical properties:	White crystal
Sources:	*Sophora flavescens*
Therapeutic effects:	Antitumor, anti-inflammation
Chemical structure:	

Name:	Prolithospermic acid
Chemical formula:	$C_{18}H_{14}O_8$ (derivative of chlorogenic acid)
Physical properties:	White powder
Sources:	In vivo biotransformation of slavianolic acid B found in radix *Salvia miltiorrhiza* (danshen)
Therapeutic effects:	Treatment for coronary heart disease
Chemical structure:	

Name:	Resveratrol
Chemical formula:	5-[(E)-2-(4-hydroxyphenyl)ethenyl]benzene-1,3-diol, $C_{14}H_{12}O_3$
Physical properties:	White powder with slight yellow cast
Sources:	Skin of grapes, blueberries, raspberries, and mulberries
Therapeutic effects:	Anti-inflammation, anticancer
Chemical structure:	

Name:	Rosmarinic acid
Chemical formula:	(2″R″)-2-[[(2″E″)-3-(3,4-Dihydroxyphenyl)-1-oxo-2-propenyl]]oxy]-3-(3,4-dihydroxyphenyl)propanoic acid, $C_{18}H_{16}O_8$ (derivative of chlorogenic acid)
Physical properties:	Red-orange powder
Sources:	Radix *Salvia miltiorrhiza* (danshen), the fern family *Blechnaceae* and *Lamiaceae*
Therapeutic effects:	Anxiolytic and acts as a GABA transaminase inhibitor, antioxidant
Chemical structure:	

Name:	Safflomin
Chemical formula:	(6*E*)-2,5-Dihydroxy-6-[(*E*)-1-hydroxy-3-(4-hydroxyphenyl) prop-2-enylidene]-2,4-bis[(2*S*,3*R*,4*R*,5*S*,6*R*)-3,4,5-trihydroxy-6-(hydroxymethyl)oxan-2-yl]cyclohex-4-ene-1,3-dione, $C_{27}H_{32}O_{16}$
Physical properties:	Yellow pigment
Sources:	Safflower (*Carthamus tinctorius*)
Therapeutic effects:	Anticancer
Chemical structure:	

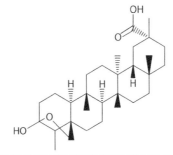

Name:	Salaspermic acid
Chemical formula:	(3*S*,5a*R*,7a*S*,7b*R*,9a*S*,12*R*,13a*R*,13b*S*,15a*R*,15b*S*,16*R*)-3-hydroxy-7b,9a,12,13b,15a,16-hexamethylicosahydro-3,5a-methanochryseno[2,1-c]oxepine-12(5*H*)-carboxylic acid, $C_{30}H_{48}O_4$
Physical properties:	White solid
Sources:	*Salacia macrosperma* and *Tripterygium wilfordii*
Therapeutic effects:	Anti-HIV (treatment of AIDS)
Chemical structure:	

Name:	Salvianolic acid b
Chemical formula:	(2R)-2-[(E)-3-[(2R,3R)-3-[1-carboxy-2-(3,4-dihydroxyphenyl) ethoxy]carbonyl-2-(3,4-dihydroxyphenyl)-7-hydroxy-2,3-dihydro-1-benzofuran-4-yl]prop-2-enoyl]oxy-3-(3,4-dihydroxyphenyl)propanoic acid, $C_{36}H_{30}O_{16}$ (derivative of chlorogenic acid)
Physical properties:	White to tan powder
Sources:	Danshen (*Salvia miltiorrhiza*)
Therapeutic effects:	Antioxidant and anticancer
Chemical structure:	

Name:	Tanshinone IIA
Chemical formula:	1,6,6-trimethyl-8,9-dihydro-7H-naphtho[1,2-g][1]benzofuran-10,11-dione, $C_{19}H_{18}O_3$
Physical properties:	Red powder
Sources:	Danshen (*Salvia miltiorrhiza*)
Therapeutic effects:	Antioxidant and anti-inflammation
Chemical structure:	

Name:	Tetrandrine
Chemical formula:	6,6',7,12-tetramethoxy-2,2'-dimethyl-1 beta-berbaman, $C_{38}H_{42}N_2O_6$
Physical properties:	Needles
Sources:	*Stephania tetrandra* S. Moore
Therapeutic effects:	Anti-inflammatory, immunologic, and antiallergenic effects
Chemical structure:	

Name:	Triptonine b
Chemical formula:	Methyl 2-(tetraacetoxy-dihydroxy-pentamethyl-pentaoxo-acetate), $C_{46}H_{49}NO_{22}$
Physical properties:	Powder
Sources:	*Tripterygium hypoglaucum*
Therapeutic effects:	Anti-HIV activity
Chemical structure:	

Index